Environmental Reform in the Information Age

T0253723

As the information revolution continues to accelerate, the environment remains high on public and political agendas around the world. These two topics are rarely connected, but information – its collection, processing, accessibility and verification – is crucial in dealing with environmental challenges such as climate change, unsustainable consumption, biodiversity conservation and waste management. The information society (encompassing entities such as the Internet, satellites, interactive television and surveillance cameras) changes the conditions and resources that are involved in environmental governance: old modes and concepts are increasingly being replaced by new, informational ones. Arthur P. J. Mol explores how the information revolution is changing the way we deal with environmental issues, to what extent and where these transformations have (and have not) taken place, and what the consequences are for democracy and power relations. This book will appeal to scholars and students of environmental studies and politics, political sociology, geography and communications studies.

ARTHUR P. J. MOL is chair and professor in environmental policy in the Department of Social Sciences at Wageningen University. He is the author of *Globalization and Environmental Reform: The Ecological Modernization of the Global Economy* (2001) and *The Refinement of Production: Ecological Modernization Theory and the Chemical Industry* (1995).

Environmental Reform in the Information Age

The Contours of Informational Governance

ARTHUR P. J. MOL
Wageningen University, The Netherlands

CAMBRIDGE
UNIVERSITY PRESS

CAMBRIDGE UNIVERSITY PRESS
Cambridge, New York, Melbourne, Madrid, Cape Town, Singapore,
São Paulo, Delhi, Dubai, Tokyo, Mexico City

Cambridge University Press
32 Avenue of the Americas, New York, NY 10013-2473, USA

www.cambridge.org
Information on this title: www.cambridge.org/9780521182652

© Arthur P. J. Mol 2008

This publication is in copyright. Subject to statutory exception
and to the provisions of relevant collective licensing agreements,
no reproduction of any part may take place without the written
permission of Cambridge University Press.

First published 2008
First paperback edition 2010

A catalog record for this publication is available from the British Library

Library of Congress Cataloging in Publication data

Mol, Arthur P. J.
Environmental reform in the information age : the contours of informational
governance / Arthur P. J. Mol.
  p. cm.
Includes bibliographical references and index.
ISBN-13: 978-0-521-88812-7 (hardback)
1. Environmental policy. 2. Environmental management.
3. Information technology. I. Title.
GE170.M6195 2008
363.7′05–dc22   2007041634

ISBN 978-0-521-88812-7 Hardback
ISBN 978-0-521-18265-2 Paperback

Cambridge University Press has no responsibility for the persistence or
accuracy of URLs for external or third-party internet websites referred to in
this publication, and does not guarantee that any content on such websites is,
or will remain, accurate or appropriate.

For Wilma, Kasper and Marente

Contents in Brief

Contents

Tables, figures and boxes

Tables

Figures

Boxes

Preface

In the beginning of 2007 – one and a half decades after the former environmental upsurge – environment is again high on the public and political agendas around the world, not in the least through Al Gore's media campaign around the Oscar-winning movie/documentary *An Inconvenient Truth*. At the same time the information revolution continues to amaze people and to change life of many of us – through Internet, through blogs, through time-space compression. But the two – environment and information – have hardly been connected. In the growing number of stories on the major environmental challenges planet earth is facing, information is not really among the exciting topics that easily find their way into the newspapers, prime-time news, or even academic literature. And the digitalization of our life, the acceleration of information flows, and the enhanced potentials of monitoring, tracking and tracing are not often related to environmental sustainability. And still, I will argue in this book, information – that is, its collection, processing, accessibility and verification – is becoming crucial in dealing with climate change, unsustainable consumption, biodiversity conservation and waste management, to name but a few. The information society is rapidly changing the conditions, mechanisms, resources, institutions and conflicts that are and will be involved in environmental governance. Old modes, resources, arrangements, concepts, and sites of power are increasingly being replaced by new, informational ones. This is not only true in the economy, where the old Fordist mode is gradually being replaced by a new informational economy, but also for environmental governance, where the dominance of national, state-based, command-and-control environmental regulation is being broken down in favour of transnational, public-private, networked, informational governance. This book explores how the information revolution is changing the way we deal with environmental challenges; to what extent and where these transformations have (and have not) taken place and what the consequences are of these new modes of environmental governance through information, for instance, for democracy, power relations and the so-called informational peripheries in the South.

Ideas and drafts for the various subjects and chapters that make up this book have been presented and discussed in a number of conferences and workshops around the globe: the international conference "Governing Environmental Flows" (Wageningen, the Netherlands, June 2003); the international conference "Technology, Risks and Uncertainty: Challenges for a Democratization of Science" (Florianapolis, Brazil, April 2004); the workshop "Globalization, Forest Governance and Forest Certification" (St Petersburg, Russia, May 2005); the international conference "Environment, Knowledge and Democracy" (Marseille, France, July 2005); the international conference "Flows and Spaces in a Globalised World" (London, UK, August/September 2005); the World Congress of the International Sociological Association, especially its Environment and Society sessions (Durban, South Africa, July 2006); the international conference "Environmental Management of Urban and Industrial Infrastructure in Asia" (Ho Chi Minh City, Vietnam, October 2005); the international conference "Globalization, Environmental Ethics and Environmental Justice" (Lanham, Michigan, August 2006); and the international conference "Greening of Agro-industries and Networks in Asia" (Bangkok, Thailand, October 2006). I feel privileged to have been able to test my ideas at so many different locations for such diverging audiences. I greatly appreciate the discussions and exchanges with participants at these various meetings, as well as with the students and staff at the Environmental Policy Group of Wageningen University, the Netherlands. Together, these audiences constructed the reflections and feedback that are essential in any maturation of ideas.

A number of my students deserve special mention for their help in writing this book: Sander van den Burg, Le Van Khoa, Pham Minh Hai, Pham Van Hoi, Liu Yi and Jinyang Zhang. My colleagues Steven Yearley, Harald Heinrichs, Pierre-Benoît Joly, Zhang Lei, the late Fred Buttel, David Sonnenfeld and – as always – Gert Spaargaren were especially helpful in sharpening my ideas and lines of argumentation through their discussions, reading of texts and constructive criticism.

Wageningen, April 2007

1 | *Introduction: new frontiers of environmental governance*

For some, like the newspaper 'The European', the Agency might be seen as 'a watch-dog without teeth'. Or as Lord Tordoff remarked recently, we could become 'just an information black hole'. For others, who are aware of the power of information in our society, the Agency could go to the opposite extreme and become a concealed power center, a Trojan Horse.

Domingo Jimenez-Beltran, executive director of the EEA, 1995

1. The dawn of a new era

Twelve years and more than sixty thousand orbits on from its launch, the Earth Observation mission of Europe's Space Association ESA, ERS-2 satellite, continues with all instruments functioning well. ERS-2 was launched on 21 April 1995, ensuring continuity of data from ERS-1, the first European Remote Sensing program mission. A growing global network of ground stations of more than three thousand users is receiving data from the veteran spacecraft ERS-2. When the Asian tsunami struck in December 2004, satellites provided rapid damage mapping. Another ERS-2 sensor working in near-real time is its Global Ozone Mapping Experiment (GOME), delivering atmospheric global coverage of ozone and other trace gases and supporting operational services such as Tropospheric Emission Monitoring Internet Service (TEMIS), which provides daily ozone, ultraviolet and air pollution monitoring. The latter functions are mainly supported by another satellite, Envisat. In March 2002, the European Space Agency launched Envisat, an advanced polar-orbiting earth observation satellite that provides measurements of the atmosphere, ocean, land and ice. The Envisat satellite has an ambitious and innovative payload that will ensure the continuity of the data measurements of the ESA European Remote Sensing satellites. Envisat data support earth science research and allow monitoring of the evolution of environmental and climatic changes. Next to the large number of professional users,

Envisat information systems are also publicly disclosed via various Web sites, providing almost real-time and strongly visualised environmental data and information.[1] Among others, its first images of air pollution formed the start of various nongovernmental organisation (NGO) campaigns in Western Europe to combat air pollution, making an almost similar impact on environmental governance as the first pictures of 'spaceship earth' in 1970 (mobilising politicians and civil society) and the first visualisation of the hole in the ozone layer in the early 1980s (pushing the Vienna Treaty negotiations and marking the offset of global environmental governance). The major differences are, however, significant: where the pictures of spaceship earth and the hole in the ozone layer were almost one-time events, carefully constructed and transmitted by the scientific community to the public via newspapers and television, the continuous flows of real-time air pollution data and visualisations of all parts of the globe are at any time available through the Internet for everybody with access to the Internet.

But it is not only high-technological and global-oriented developments in environmental information systems and arrangements that seem to mark a new era in dealing with environmental challenges. In 2002, the European Parliament and the European Council issued the European Union's General Food Law. Next to the establishment of a European Food Safety Authority, the law requires a far-reaching system of tracking and tracing to be developed. Food products need to be tracked in their movement through food commodity chains and networks, from the farmer or even animal to the final shop or consumer. In addition, it should be possible, according to the European Union (EU), to trace the origin of food products at retailers and consumers. This resulted in ongoing efforts to develop information systems that allow the tracking and tracing of food and feed, but increasingly also of nonagricultural products, through commodity networks. Although to a significant extent the food safety crises in the EU and beyond were at the origin of government demand for these sophisticated tracking and tracing systems, increasingly issues of transparency, social and environmental responsibilities, product stewardship and product life cycle analyses and consumer trust require advanced systems of data and information collection, handling, transmission and application beyond governmental agencies. This resulted not only in new

---

[1] See, for instance, http://envisat.esa.int/ or http://www.temis.nl/ for almost real-time images of various air pollution indicators in all parts of the world.

information systems, but also in new time-space connections between producers and consumers; new governance arrangements between private and public actors; the emergence of auditing, inspection and verification agencies and new questions of accountability, transparency and trust. But do these new institutional arrangements form an adequate answer to the uncertainties, risks and consumer anxiety that seem to have become part of the globalised food system? Are these informational arrangements better able to deal with those questions and doubt than the conventional scientific and nation-state institutions?

These two examples seem to mark a new era in the role of information and informational processes in governing the environment. It is not just the (supranational or global) scale of information collection, handling, spreading and use that point to these innovations, but also the sheer amount of information, the speed of information processing, the availability of information for ever-wider groups in society and the growing importance (or power) of information resources in environmental struggles that contribute to that. But do such innovations in environmental knowledge and information collection and handling really mark a new way of how modern society approaches its environmental challenges? Are the Envisat and the tracking and tracing system not just one further small step in an ongoing development of collecting information for governing the environment, a development that started at the birth of modern environmental policy in the 1960s and will be continuously refined incrementally? This chapter sets the argument that we have to rethink the role of information in environmental governance – that our present world witnesses a qualitatively new phase in the relation between information processing and environmental governance. That new phase is partly – but not primarily – a product of new technological advancements that enlarge our informational capabilities.[2] But it is, moreover, marked by a number of wider developments

---

[2] There have been earlier revolutionary developments in information and communication technologies (such as telegraph and the telephone system in the nineteenth and early twentieth century), which had a major impact on modern society's economic and social life. These also affected the 'old' social movements, in terms of increased speed and range of communication (cf. Tarrow, 1998). But these earlier communication innovations have been hardly relevant for changes in environmental governance, as around those days environmental protection was hardly developed and articulated in a full-fledged relative autonomous subsystem in modern society, with its own rationalities, institutions and organisations.

in global modernity, which strongly condition and structure new modes of environmental governance in which information and informational processes become crucial elements. And with this growing importance of information and informational processes new questions emerge with respect to environmental governance: for instance, questions related to access to and control over information and informational processes; questions related to quality, reliability, uncertainty and verification of information; questions related to new power relations between non-state actors and state authorities and questions related to new institutional arrangements to govern the environment in an era marked by information centrality. It is these kinds of questions that this book aims to address.

2. Information explosions

Knowledge and information on the environment have been of crucial relevance for environmental policy making, governance, and reform ever since Rachel Carson (1962) started a new wave of environmental concern and reform with her path-breaking work on pesticides. By revealing how pesticides in agriculture accumulated in food chains and endangered natural ecosystems and human health, Carson not only gave scientific proof of their toxicity but also started a public campaign that put environmental side effects of (simple) modernisation strongly on the public and political agendas. Environmental information, and especially natural science – based knowledge and information on the natural environment, has been – and continues to be – an important factor in designing environmental reform measures and strategies. It is dazzling to imagine the amount of environmental data, information, and knowledge (cf. Box 1.1) being collected almost in a routine way on a daily basis through environmental examinations of air and water quality, through state-of-the-environment reporting programs; through information gathering on species, ecological systems and their vitality; through inspections on toxic substances in food and other products; through emission monitoring of companies and farms; through the domestic metering of water, electricity and even waste flows in and out of households and so on. The EU provides even a wider definition of environmental information. Environmental information then refers not only to information on the state of the environment, or on the 'additions and withdrawals' (the emissions and exploitation of natural resources). The recently adopted EU Directive 2003/4/EC on public

Box 1.1 Clarification on terminology

Throughout this book, I will use the concept of information as the overall category, rather than data or knowledge. During the various conferences in which ideas on this book have been presented, several scholars have questioned information as the central concept, often preferring knowledge as the key category. There are several reasons not to do so. This work is especially in line with debates on the Information Society and the Information Age, and less in the tradition of constructivism, the sociology of science, expert knowledge versus lay knowledge and so on. Although the concept of Knowledge Society is becoming slowly common (e.g., Stehr, 1994; UNESCO, 2005), Information Society and Information Age are more widely used in these debates. Second, knowledge refers to processes, problems and struggles on interpretation (through science or other frames) of information. Although that is definitely relevant and will emerge throughout the book, this is too limited for understanding information-related changes in environmental governance. Equally relevant for our analysis are the digitalisation of information, the information and communication technologies, the time-space compression in information circulation and flows, and so on, which are all referring to information rather than to knowledge. In that sense, I interpret information as a somewhat more general category than knowledge. As for the distinction between information and data: Jimenez-Beltran (1995), executive director of the European Environmental Agency (EEA) at that time, makes a distinction between environmental data, which refers to – often quantified – numbers and figures on environmental conditions, and environmental information, which points to meaningful flows of signs for a targeted audience. Usually information refers to raw data that are processed, selected and translated to address meaningfully an audience. The key problem that Europe faces in today's Information Society, according to Jimenez-Beltran, is the contradiction between data abundance and information scarcity. Esty (2004) makes a similar distinction: 'Data is the raw material. Information is the intermediate good, reflecting some processing of the data. Knowledge is the final product where analysis allows us to extract conclusions'. I will not follow this distinction strictly, but will use information as a common overall denominator.

access to environmental information (*OJ* L 041, 14/02/2003) defines in article 2 environmental information as written, visual, aural, electronic and other material forms of information: (i) on the state of the environment; (ii) the factors, emissions and withdrawals influencing the state of the environment; (iii) environmental measures and policies; (iv) reports on the implementation, cost-benefit and other economic analysis; (v) the state of human health and safety, including food chains, built structures, cultural values and so on.

Initially, during the 1960s and 1970s, environmental information collection and handling was primarily a state task, and environmental information also was primarily used – or meant to be used – by state authorities in protecting the environment. State agencies relied, of course, on scientific institutes to do part of the data collection, monitoring and reporting, often within state-run programs. Subsequently, other nonstate actors started to get involved in the collection and handling of environmental information. Environmental nongovernmental organisations began their own programs of information gathering and knowledge building to countervail the information monopoly of the economic and political centres. Later on, private economic sectors became actively involved in environmental monitoring and information collection, either forced by state regulation, or more 'voluntary' for internal purposes (better management of environmental and natural resource flows to save money or increase product quality) or external reasons (collecting countervailing evidence against NGO pressure, setting up public relation campaigns, building annual environmental reports, fulfilling requirements set by customers, preventing legitimacy questions). The diversification of information collection and handling agencies contributed to the enhancement of environmental data and information, making information increasingly available for larger groups in shorter time periods at more and more locations around the globe.

Among these numerous actors and institutions involved in information generation, collection, handling and distribution science and scientists have played a particular role. It is not only that science and scientists have been crucial with respect to the generation of new information and knowledge on cause-effect relations of substances released in the environment; the development of new environmental measuring, monitoring, data storage and data analysis technologies; the compilation of state-of-the-environment reports and advancements on modelling and prediction, among others. Arguably more of importance, science (and natural scientists) was for a long time seen as a landmark to assess and distinguish true from false information, public relations from disinterested dissemination, balanced judgements from self-interested biases and apocalyptic predictions from comforting naivety. Although we will see that such an 'arbiter view' of science is no longer adequate today, it did help the rapid and further institutionalisation of environmental sciences and studies in academia, until it gained a comfortable established position by the end of the 1980s. Environmental institutes, university departments, course and education programs, academic journals and book series are the institutionalised witnesses. It does not seem that we are about to reach the finish of the rising natural science production of environmental information and knowledge.

But although the amount of available environmental knowledge and information is growing on almost all environmental issues for all kinds of decision makers (private and public, institutional and individual) through increasing scientific research, monitoring practices, information storage capacity and high-speed and long-distance information transport, reflections and interpretations on what these developments in environmental information mean for the way modern society handles the environment have been rather poor. After summarising how the environmental social sciences have conventionally studied environmental information (Section 3), I will set the stage for this book – and introduce its various chapters and themes – by arguing that a new Information Society/Information Age perspective needs to be developed to understand how contemporary society develops new informational modes in dealing with environmental challenges.

3. Conventional interpretations of environmental information

Historically, three social science research traditions have explicitly focused on interpreting environmental information and knowledge processes with respect to issues of environmental governance: a more conventional tradition following attitude-behaviour models; a more policy/economics/legal tradition focusing on information gaps, distortion and transaction costs and a more critical one based on constructivism and the Risk Society thesis.

The first two traditions – established in the late 1960s and 1970s, but still applied today – focus on environmental science, knowledge and information in dealing with environmental crisis in a rather straightforward – we would now say: simple modernity – way. The first approach focuses on information dissemination following a basic attitude-behaviour model (in line with the original ideas of Fishbein and Ajzen, 1975). Basically, environmental information is considered essential for changing individual environmental attitudes, and attitudes are believed to be fundamental for behavioural changes (with respect to for instance mobility, consumption, waste handling). The spreading of environmental information on causes, consequences and alternative behaviours is often linked to voluntary programs that invite polluters to rearrange their daily routines of production and consumption into more sustainable directions (cf. Moxen and McCulloch, 1999).[3]

The second, more policy-oriented perspective has a more legal and economics background and emphasises the necessity of producing and collecting knowledge and information by environmental experts and authorities to develop a solid – natural science – basis for environmental policies and reform. Starting from the 1970s information gaps, transactions costs and private ownership of information have been identified as distortions of effective and efficient environmental policy making, implementation and control. Information is seen as of crucial importance for overcoming regulatory failures, and according to these scholars better information, more information and cheaper information improve environmental decision making.[4] Environmental governance is then seen as solidly based on expert knowledge and information, strongly in line with the conventional rationalist policy theories and pluralist state theories (cf. Ham and Hill, 1984). Nthunya (2002) and Burström and Lindqvist (2002) are recent examples in this perspective. Recent ideas on and quests for data-driven policy making to bring rationality in decision-making processes (cf. Esty and Rushing, 2006) are modern, intelligent versions of the same school.

Both schools within this tradition interpret environmental knowledge and information rather straightforwardly, as unproblematic categories, in which more knowledge and information is positively and causally related to better environmental governance. The application

---

[3] Arguably, the analyses of the use of information by environmental NGOs in their media strategies and campaigns are among the relatively more 'sophisticated' studies in these traditions.

[4] Esty (2004) provides a full overview of the literature on information obstructions and information needs in environmental policy-making processes, ranging from problem identifications to enforcement and evaluation.

of environmental information by various kinds of state and nonstate actors is not really problematised; only the lack of information, the information gap, misleading information, poor dissemination and failing interests are criticized.

In the 1980s, a third, critical perspective emerged on environmental information. It has been especially the work of social constructivists and the studies related to Ulrich Beck's (1986) Risk Society hypothesis that brought environmental social science scholars to critically focus on the role of science and information in environmental crises. Scholars in this tradition investigated the changing role of science, scientists, and experts, and scientific information in social practices and institutions, focusing on the loss of authority of and growing ambivalence towards scientific knowledge and information. Discussions of the role of scientists, science and scientific knowledge in decision-making processes flourished already for quite some time but basically internally, within the scientific domain. It was the opening of these discussions and ambivalences to wider domains in society that resulted in increasing (feelings of) uncertainty among decision makers and lay actors in society. The recent discussions with respect to Lomborg's (1998) study *The Skeptical Environmentalist*, but also discussions on genetically modified organisms (GMOs), food crises and climate change, seem to give evidence and underline the uncertainties some claim to be inherent in today's scientific studies, monitoring practices and measurement and abatement strategies on the environment. In addition to governmental decision makers at all levels facing conflicting interpretations and uncertainties, citizens and consumers are almost on a daily basis confronted with contrasting claims with respect to the environmental or health consequences of products and social practices, without having one clear authority that sifts true from false information and claims. This critical perspective questions the standard idea that scientific insights, knowledge production and information collection and dissemination contributes to better environmental governance and reform. More scientific information, according to these critics, contributes rather to more debate, uncertainties and doubts on problem definitions, abatement strategies and environmental risks. Is more environmental knowledge and information paralysing environmental protection and reform? Or, to put it more sociologically, are we confronted today with only "regressive uncertainty so that the more we know, the more uncertainty grows" (Urry, 2004: 10)?

Consequently, information and knowledge became inherently contested and could/should be deconstructed, contributing to and emphasising the disputable contribution of scientific information to environmental reform. In their detailed, sophisticated and innovative analyses on the changing role of knowledge and information in dealing with environmental challenges, scholars in this tradition were rather one-sided in neglecting to a large extent the formative role of information and knowledge in environmental governance and reform (so central in the first two traditions). In addition, their alternative was not so much less information and less knowledge, but a better understanding of the limitations of scientific knowledge (especially versus lay knowledge) and more insights in the politics and power of knowledge generation and dissemination.

Although all three traditions are far from outdated and continue to produce rich and valuable insights, they miss – each in their own way – the overall perspective, tools and focus to understand and interpret how the contemporary information *explosion* and information *revolution* are transforming the way modern societies (try to) cope with their environmental challenges. Because their origins and backgrounds are in the 1970s and 1980s, this is not too surprising. A starting point for a more overarching perspective needs to be founded on ideas and perspectives of the Information Society and the Information Age (but cannot remain limited to those for their failure to take the environment into account, as we will argue next).

4. The Information Society and the missing environment

Ideas on the growing role of information in the construction and transformation of modern society date back to the late 1960s and 1970s when a first group of authors started to reflect on what was then commonly referred to as the postindustrial society or the Information Society.[5] "The post-industrial society is an information society, as industrial society is a goods-producing society" (Bell, 1973:467). Although initially the basic idea of the postindustrial society was the movement to a service society and information itself remained relatively undeveloped, in the 1970s information moved to a more central position: "My

---

[5] The notion of Information Society emerged for the first time more systematically in Japan in the late 1960s. Among Japanese scholars a major interest has remained in the idea of the Information Society (Kumar, 1995).

basic premise has been that knowledge and information are becoming the strategic resource and transforming agent of the post-industrial society" (Bell, 1980: 531). Social scientists such as Daniel Bell (1973, 1980), Alvin Toffler (1970, 1981) but also Alain Touraine (1971) and even André Gorz (1982) started to reorient their initial writings of the material underpinnings of modernity to more cultural processes, service sectors, new social movements and information. Although these and other scholars share some basic starting ideas – most notably for us the growing importance of information and informational processes in social development – it would be a major mistake to put all postindustrial society and Information Society scholars and advocates under one common denominator. In an illuminating study, Boris Frankel (1987) analysed the similarities and differences in what he labelled left-wing and right-wing postindustrial utopians. In analysing the origins, continuities and differences of postindustrial thinkers and postmodernism, Kumar (1995) equally illustrated the variety of ideas and analyses among postindustrial and Information Society analysts and advocates. And, finally, in a number of books, Frank Webster (2002; Webster and colleagues, 2004) has brought together the main founding fathers and contributors to the Information Society thesis, illustrating their diverging approaches, assessments and theoretical stance.

In academic studies on the changing character of modern society in the 1980s information lost its prominent place, only to regain it in the 1990s with the emergence of a group of scholars that we will bring together under the label Information Age.[6] According to these scholars, in the 1990s information, information technologies and informational processes became crucial in understanding and defining a new phase in modernity, often labelled late, reflexive or global modernity. Especially with the work of Manuel Castells (and especially his major trilogy *The Information Age*; 1996/1997), information became widely accepted as a key element that restructures modern society under conditions of globalisation. With him, general and theoretical sociologists such as Anthony Giddens, Ulrich Beck and John Urry[7] focused on information and communication technology, globalisation and global networks and

---

[6] Several authors earlier – before the 1990s – coined the term Information Age (e.g., Naisbitt, 1984, for a timely use).

[7] Others have preceded or followed their line of investigating (cf. Sassen, 1994; Gunaratne, 2002), resulting in what we have labelled elsewhere a sociology of networks and flows (Mol and Spaargaren, 2006).

flows as crucial and essential architectures of a new phase of modernity. Knowledge and information, and its networks and infrastructures, were increasingly interpreted as (one of) the essential categories and formative factors in entering the new millennium.

The thesis of an emerging Information Society or an Information Age has been debated heavily, and Chapter 2 goes into the details of these ambivalences and debates. In this introductory chapter, I want to stress especially the missing environment in the literature and debates on the Information Society/Age. Quite surprisingly, in their analyses of a changing modern society through information, information and communications technology (ICT), and informational processes all major social theorists have almost neglected the environmental domain. Initially, in the writings on the postindustrial society and the Information Society, the environment figured marginally as one of the driving forces and positive consequences of social change. Also, a few of the more recent studies in the Information Age literature use incidental examples and illustrations from the physical environment in their argument for the transformations that can be witnessed (cf. Giddens, 1990; McNaghten and Urry, 1998; Beck in various writings). But the major scholars of the Information Society and the Information Age have not really included the environment fully in their analyses, as will be argued and illustrated more in depth in Chapter 2.

5. Environmental assessments of the information revolution

With the 'information revolution' in the 1990s, the widespread emergence of information and communication technologies around the globe and the growing centrality of information in all kinds of social, political and economic processes, new research traditions on environment and information are emerging. We can distinguish three lines of analysis, where academics try to assess the consequences of these new informational developments for the environment. Although social scientists do play a role in such analyses, the problem definition is more than incidentally of a technical nature.

A first line of analyses seem to concentrate on the direct environmental implications of these new technological systems related to information. This line falls apart in scholars stressing the potential environmental dangers of the ICT revolution, and those – much in line of the glorifiers of the postindustrial society – who celebrate the positive environmental outcomes. So, the environmental side effects of computers and IT industries are a popular field of investigation. Plepys (2002) and Tulbure (2002), for instance, discuss under the concept of rebound effects the environmental consequences of increasing consumption and production of ICT products and services, relating these consequences to both the life cycle of ICT products and the way these products are used. In a similar way, Berg (2003) argues that the concept of a paper-less office is rebounded by the fact that through ICT a written office let-ter is printed out eight times on average before it is sent. Romm (1999) assesses the energy use of the Information Age economy and, hence, its effect on global warming. Smith, Sonnenfeld and Pellow (2006) bring together numerous studies on what the production of ICT equipment does to labour, health and the environment in a large number of coun-tries throughout the world. More general evaluations on the ecology of the new economy are given by Park and Roome (2002), Grubler et al. (2002) and Brown (2001). The work of Heinonen et al. (2001) is an example of a borderline study. They start their analysis with the environmental side effects of information technology, but turn this into preventive action in developing the idea of a sustainable Information Society, referring to the necessity and possibilities of greening the ICT sector. At the other side of the same research paradigm we can iden-tify analyses of the contribution of ICT to dematerialisation, imma-terialisation and sustainable development, for instance, by lowering paper use (often not being proved); reducing travelling behaviour as a result of distant learning, teleconferencing, teleworking and commut-ing and the like; fine-tuning production processes or replacing mate-rial consumption by virtual consumption (for instance, in tourism). Dematerialisation is than conceived as the progressive gains in eco-efficiency, especially via improvements in production; immaterialisa-tion is defined as a radical sudden shift towards nonmaterial products to fulfil human needs initially fulfilled by material products, especially in the domain of consumption (Simmons, 2002). Although the sustain-ability prospects of ICT (or Information Society Technologies IST, as the EU prefers to call them in their sixth R&D framework program) are usually considered large, the actual achievements in terms of envi-ronmental improvements are still modest. A special issue of *The Jour-nal of Industrial Ecology*[8] completely focused on these direct negative

---

[8] Volume 6 (2002), no. 2, has papers that dive into the environmental effects of ICT in sectors as diverse as travelling, e-commerce, the retail sector, the media and the energy sector.

and positive environmental side effects of ICT and its application. The TERRA 2000 research project, part of the sixth EU framework program, equally looked into and tried to assess the potential (future) contribution of Information Society technologies to sustainable development.[9] Studies of Salzman (1999) and Forseback (2000) and analyses such as those of Slob and van Lieshout (2002) are in the same tradition. A recent EU study by Erdmann and colleagues (2004) assesses impacts between minus 20 percent and plus 30 percent in environmental deterioration/improvement by 2020, due to ICT.

These studies find their limitations in predominantly a technological definition of the information revolution, although some studies (such as the one of Slob and van Lieshout, 2002) include organisational dynamics and consequences. What is new today, compared to say the 1980s, is basically another technological profile of modern society, and we have to assess what the direct and indirect environmental consequences in terms of 'additions and withdrawals' are – and can be – of these new technologies, technological systems and technological potentials. The typical questions of this research paradigm are: Do these new technologies produce more or less waste, do the make production more or less environmentally efficient, can they replace material needs for nonmaterial needs, do they lead to less or more resource extraction and emissions?

Few environmental studies on the information revolution broaden the scope a little further, beyond the direct environmental effects of ICT and information systems. In these broader studies, there is a second group of studies that can be labelled 'nothing new to report'. Most of these see the information revolution, and the information technology linked to that, as nothing really new: it is just another phase of global capitalism, in which ICT might reorganise production and consumption processes in new time-space constellations but not into fundamentally different environmental profiles, consequences and processes. E-business, flexible network companies, global capital flows, a worldwide telecommunication network might all restructure the global economy, but they leave the material underpinnings as well and the fundamental power imbalances largely unaltered. Nothing fundamentally

---

[9] This research project ran from 1998 to 2003 and was carried out by various national research teams in Europe. For the various outputs of this project, as well as its final report, see http://www.terra-2000.org.

new to report. This is supported by economic scholars who try to quantify the 'weightless economy', concluding that the Internet economy or e-economy is still so small in terms of global GDP and so localised especially in the United States that it does not make any serious environmental impact in terms of dematerialisation (cf. UNCTAD, 2001; Thompson, 2004). It also builds on existing environmental research on the role of conventional media such as television, journals and radio, focusing strongly on media coverage of environmental problems, the framing of environmental issues, the power of media conglomerates and elites in information control, as well as more instrumental research of effective information dissemination. Although now the conventional information technologies are extended to new ones (ICT, Internet, mobile phones), the kind of research, the theoretical schemes and the conclusions are not really different

Finally, there is a third group of scholars who does start from the idea that the information revolution does change the way we deal with the environment: how environmental governance take place, how information is reorganising various social processes that have relevance for the environment, how new information processes change the power balances and resource distribution within societies and across the globe and so on. Although this group of scholars will be our starting point for this book, their numbers are limited. A good example is formed by Heinonen, Jokinen and Kaivo-oja (2001), who pay attention to the increased possibilities of the information revolution for communicating environmental information through production – consumption chains, paying attention to especially all kind of technological devices that enable actors to identify the (environmental) characteristics of (half) products and use that information in more environmentally rational behaviour. Others focus at electronic regulation (or e-governance) on environmental issues; at new possibilities for environmental transparency, accountability and legitimacy; at lower transaction costs, larger data availability, decreased uncertainties and increased monitoring potentials through the digital technologies (e.g., Esty, 2004); or at the enhanced possibilities for environmental activism. These perspectives and ideas are complemented by other scholars who emphasise the drawbacks of the information revolution on environmental governance: the overkill of information, the growing and structural uncertainties that paralyse environmental state authorities as well as citizen-consumers in taking environmental action, the new (digital) divides in international environmental governance and the strategic misuse of environmental information. It is exactly these contrasting ideas – these two sides of the same coin – that construct the building stones for developing a new environmental governance perspective, what I will label informational governance on the environment. Such a perspective tries to grasp how, in which way and to what extent the Information Society/Age is changing the social processes and dynamics of environmental governance and reform.

6. Shifting (environmental) governance

The emergence of informational governance should be understood in relation to the relatively recent debate on (modes of) governance. At the turn of the millennium, a debate on new forms and modes of governance emerged especially in the political sciences (and to a lesser extent in other social sciences) in the industrialised world. Although certainly not restricted to the environmental domain, the governance debate often uses environmental examples for illustrating the contemporary changes that are believed to be taking place in governing practices.

In summarising the different modes of governance, the NEWGOV (2004) project comes with a classification in which the actors involved and the steering modes form the two crucial dimensions that determine the seven modes of governance, as summarised in Table 1.1. In reviewing the complex debate on (new)[10] modes of governance Treib et al. (2007) make a differentiation in three dimensions of governance that helps to understand the variations in modes of governance: politics, polity and policy. Within politics, modes of governance differ with respect to the degree of involvement of private actors. Within policy, modes of governance differ with respect to legal bindingness versus soft law, rigid versus flexible approaches to implementation, the prevalence or absence of sanctions, procedural rather than material regulation and malleable instead of fixed norms. And within polity, modes of governance can be distinguished according to market structure and dynamics vis-à-vis hierarchy, a central versus a dispersed

---

[10] Treib et al. (2007) consider the labels 'old' and 'new' not very useful as what is old in one field or domain of governance might be new in another. Instead, they prefer analytical categories and distinctions to sort out and categorize different modes of governance. Although this is correct and useful, one cannot deny that there does exist a temporality in the different modes of environmental governance.

Table 1.1. *Modes of governance* (according to NEWGOV, 2004)

		Actors involved	
	Public actors only	Public and private actors	Private actors only
Steering modes	Hierarchical, Top-down, Legal sanctions	Traditional nation-states; Supranational institutions		
	Nonhierarchical, Bargaining, Positive incentives	Intergovernmental bargaining	Delegation of public functions to private actors; Neocorporatism	Private-interest government
	Nonhierarchical, Nonmanipulative, Persuasion, Learning and arguing, Diffusion	Institutional problem solving across levels; European agencies	Public-private networks benchmarking	Private-private partnerships (NGOs)

locus of authority and institutionalised versus less formally institutionalised interactions. Although the two categorisations are somewhat different, to a large extent similar developments and categories can be witnessed.

These developments and categories are in a similar way reflected in other overviews and studies in the governance tradition, such as those of Héritier (2002), Kooiman (2003), van Kersbergen and van Waarden (2004), Pierre and Peters (2005), and Jordan and Schout (2006). Most of such studies in the governance tradition notice, describe and classify new forms of governing societal sectors and problems. Common among these authors are observations on (i) the growing involvement and power of nongovernmental actors in an increasing number of governing arrangements; (ii) the diversification of the modes of governing, diverting from a monopoly of law-based regulatory intervention towards a plurality of approaches and steering modes and (iii) the interdependencies of different levels of governance, ranging from local via national and European to truly global, and the complexities coming along with that. In explaining the diversification in forms and modes of governance the literature is less elaborated, mostly referring to processes of globalisation (in its various dimensions), growing complexities and uncertainties and the changing authority of science and the nation-state institutions.

Studies on changing modes of environmental governance have been at the foundation of, as well as built on, this more general literature on shifts in governance. Building on early studies of state failure in environmental protection (e.g., Jänicke, 1986), and following ideas of political modernisation (in the EU) or reinventing government (in the North American continent), environmental social scientists have been timely in calling for and noticing shifts in environmental governance. Lemos and Agrawal (2006) provide an extensive overview of the changes in environmental governance analysed and described in the environmental social science literature, focusing on four dimensions: globalisation, decentralisation, market-based governance and cross-scale (or multilevel) governance. Again, similar categories and dimensions of change emerge: new actors, nonhierarchical and nonlegal steering modes, flexibility, dispersed locus of authority, less institutionalised interactions and so on. Chapter 3 further reviews and investigates the literature and debates on the shifts in environmental governance and environmental reform, and thus constructs – after Chapter 2 on the Information

Society/Age – the second background chapter for developing the idea of informational governance.

Informational governance

In the extensive literature on shifts in (environmental) governance, the Information Age is largely missing; not unlike the missing environment in the Information Society/Age literature. Daniel Esty (2004) is one of the few scholars noticing that " . . . the advance of the information age has shifted our environmental protection 'possibility frontier' and opened the door to a new era of pollution control and natural resource management". It is exactly here that this book makes its contribution, both to the studies and insights on the Information Society/Age as well as to the growing empirical evidence, literature and theories on shifts in (environmental) governance. The new Information Age, with its new technological paradigm and new social organisation, has potentially far-reaching consequences for how modern society deals with its environmental challenges. Or, to put it slightly different as a research question: How, in what way and with what effects are systems, arrangements and practices of environmental governance fundamentally changing under conditions of – and through – the Information Age? Our hypothesis will be that with the coming of the Information Age, information can no longer be interpreted as just one of the – many – factors that assists and enables governmental and nongovernmental actors in designing and implementing environmental reform programs and measures. With the Information Age, information is becoming a crucial, causal and formative resource, but also a new battlefield, for new modes of environmental governance, to be labelled informational governance. Informational governance stretches far beyond the conventional paradigms of environmental state regulation of the 1970s and 1980s, and with that it shares many of the ideas of the (environmental) governance literature that emerged from the mid-1990s onward. But it significantly adds to that literature, in emphasising the crucial role of information in these shifts in governance, to be coined informational governance. In Chapter 4, I elaborate on the notion and perspective of informational governance, and the various (theoretical) debates that come along with that.

It now becomes clear that if we want to understand how the 'information revolution' affects environmental governance we have to look beyond the conventional information themes and approaches that have prevailed in the environmental social sciences until now. We cannot limit ourselves to social constructivism, to attitude-behaviour models, to better informed policy models or to the direct environmental benefits and costs of the ICT revolution. Nor can we just rely on the new governance literature, as introduced earlier. The crucial environment-related aspects of the Information Age, of ICT systems, of global information flows, are to be found beyond these research traditions. The Information Age poses an entire new set of themes and questions on environmental governance, which the Information Age literature and the governance literature have only partly addressed. Such new questions and themes include, among others, issues of

- growing uncertainties that come along with growing availability of increasing amounts of environmental information for an ever-wider community;
- growing vulnerabilities of polluters in terms of blaming their unsustainable production practices and products by governmental, market and civil society actors;
- enhanced possibilities of monitoring the environmental consequences of practices over ever larger distances in shorter time spans; and not just by states but also by local communities, households, private companies and international organisations;
- new questions of surveillance and countersurveillance now that potentials for information and monitoring skyrocket;
- new questions of power and democracy in environmental struggles, now that information, information technologies and uncertainties have become so central: Who has access to information, who owns information, who controls and governs information flows, who verifies and certifies information, who is able to build trust related to information packages?
- the relation between these new informational modes of environmental governance and the conventional state regulation on the environment: the effectiveness of environmental governance 'through' information, the distributional dimensions of informational governance (both within one country, as across countries) and the materialisation of such governance in different settings, practices and institutional arrangements and designs.

It is exactly these kinds of issues, themes and questions that form a central lead throughout the second part of this volume: Chapters 5 to 9. As informational governance is still very much in the making (or very much a promise, as some might claim), not all questions can be answered fully and with full evidence. In many cases, we will only be able to sketch the contours of what the Information Age will mean for environmental monitoring programs, for the role of states and government authorities, for environmental NGOs and their strategies, for private environmental governance and the involvement of private sectors in environmental struggles and reforms; or in short: for new forms of environmental governance. As such, informational governance explains the emergence and working of nonhierarchical, nonlegal and noneconomic modes of governance, by focusing on the growing centrality of informational processes and resources. In doing so, at the same time it repairs the information omission prevailing in most governance literature.

7. Information-poor environments

If we are to investigate what the Information Age means for shifts in environmental governance, we are most likely to find such changes and innovations within the informational centres of global modernity. Hence, the emphasis in analysing informational governance developments in this volume will be on the informational nodes, hubs and highways of our Information Society, or, in other words, within information-rich environments. It goes without saying that within the different (geographical) settings of the Information Society these informational modes diverge somewhat, as a result of, among others, the (national) cultural backgrounds, the policy styles and culture and the economic structure. Finland with its advanced informational technologies, the United States with transnational commercial empires dominating the media, the Netherlands with its relatively good access of nongovernmental organisations (NGOs) to environmental governance and Japan with a not too transparent policy culture will all have their particularities in informational governance. Several of these particularities among countries of the Organisation for Economic Cooperation and Development (OECD) will pass when we analyse the emergence of informational governance along the informational highway. Although

Table 1.2. *The information divide*

	Daily newspapers per 1000 inh. (2000)	Radios per 1000 inh. (2001)	Television sets per 1000 inh. (2002)	Personal computers per 1000 inh. (2002)	Internet users per 1000 inh. (2002)
Low-income countries	40	139	91	7.5	10
Middle-income countries	123*	360	326	45.4	80
High-income countries	284	1266	735	466.9	364

* for upper-middle-income countries
Source: World Development Report 2004.

different on various cultural, economic and political indicators, these OECD countries are still well linked to the informational networks of the global economy. But what does informational governance mean for what we could label information-poor environments?

The notion of information-poor environments points at a variety of situations where environmental information, and the informational processes linked to that are marginalised or suppressed. We can distinguish at least four ideal-typical forms of information-poor environments, knowing that in reality these ideal types mix. First, we have information-poor environments driven by economic constraints (see Table 1.2 for some data). Economic limitations then restrict informational processes, whether it be collection of information, processing of information, dissemination of information or access to information by major groups in society. The digital divide falls often in this category, but also shortages in monitoring programs and the limitations in data processing and information publication and dissemination may relate to economic shortcomings. Second, we have information-poor environments caused by political constraints. Here political factors and dynamics cause limitations in information collection, processing, spreading, access and use. Undemocratic regimes, limitations for independent NGOs to produce countervailing evidence, manipulations of search machines on the Internet and limitations on information disclosure (such as those following 9/11) are all examples in which information collection and informational processes are hindered or blocked following political reasons. Third, information-poor environments can relate to poor organisational-institutional conditions and environmental capacities. Lack of monitoring infrastructure and capacity, a poorly structured information processing institution, lack of standardisation in information collection and publication and poor information dissemination structures can all contribute to an information-poor environment, whereas the economic conditions and political structures are favourable for informational processes. Fourth, information-poor environments also can relate to problems in the cultural or interpretation frames of information, or to conflicting cultural or interpretation frames, limiting the meaning attached to and impact of information flows in specific practices and institutional arrangements. Environmental awareness and sensitivity, cultural values or dominant interpretation frames can make high levels of information flows become irrelevant for actors. This is especially so in those situations in which the information producers and/or disseminators are detached from those who (need to) have access to and consume/use information in social practices.

These four modes of information-poor environments can be distinguished analytically, but often will be mixed in reality. In developing countries, we will often find mixtures of economic, political, cultural and/or institutional causes of information-poor environments. Some of these causes behind information-poor environments also can be found in the geographies and spaces that are closely linked to the global informational networks and flows, but often the consequences will be less severe and extreme. Although information-poor environments cannot be equalised with developing countries,[11] it is most likely that (locations and practices in) such countries will face informational deficits and show characteristics of informational peripheries. In investigating informational governance with respect to the environment in

---

[11] Information-poor environments are not equal to developing countries for two reasons: (i) such environments also can be found in developed countries, and (ii) the country category is a too-specific and not always adequate unit of analysis. With respect to the latter, Beck (2004) would accuse us of methodological nationalism if we would take the nation-state automatically as the only unit of analysis of processes that are principally crossing the nation-state. As much as space-based territories, flow-based networks can be relevant categories of information poor-environments for identifying and analysing limitations of informational governance.

information-poor environments we will restrict ourselves in Chapter 10 to two countries – China and Vietnam – to explore how informational peripheries are affected by the new modes of informational environmental governance. We do that with the knowledge that it is only one empirical category of informational peripheries, and that the two countries are by no means representative for the wider set of developing countries. But by selecting these two countries we are able to grasp at least three dimensions of information-poor environment: economic constraints in information flows and processes, such as the digital divide (as a result of the developmental features of Vietnam and major parts of China), political constraints in information flows and processes (as a result of, among others, limits in the democratic character of China and Vietnam, the restricted freedom of information and the limited possibilities of environmental NGOs) and organisational-institutional constraints in information collection, handling, processing and dissemination (to some extent related to the former two constraints, but also a factor on its own, certainly in Vietnam). One can expect that China and Vietnam share with many sub-Saharan countries the economic constraints in information flows and information processes, whereas the two countries share with several other Asian countries the political constraints, and with many other countries the organisational-institutional constraints. So, although the two countries are not by themselves representative for other countries, we will most likely find here distortions in informational processes that are typical of many other informational peripheries. Hence, these case studies of informational governance in informational peripheries have relevance beyond their territorial boundaries.

8. Design and outline

Against the background, debates and developments sketched in this chapter, it should come as no surprise that this book is explorative and agenda-setting, rather than evaluative and conclusive. It will not give any final conclusions on informational governance, but, rather, opens up a new perspective of current developments and a new research agenda on environmental governance. With these new informational developments just beginning and a research orientation only in the making it is too early for any final conclusions. At the same time, with informational governance still very much in the making, we are also able to (re)direct these new developments, caution for the unwanted (side) effects, warn against the losses of the conventional environmental governance systems, point at the inequities and inequalities that (can) come along with such a new governance mode and call for change.

As introduced earlier, the book is organised in two main parts. Chapters 2 to 4 make up the theoretical part of this book, whereas the Chapters 5 to 10 deal with different substantive aspects of informational governance.

Chapters 2 and 3 are complementary theoretical chapters. In Chapter 2, I review the literature on the Information Society and the Information Age in search for their contribution to understanding how information, information processes and information technologies (can) impact environmental governance and reform. In Chapter 3, I work in a reversed way: by reviewing the literature on environmental governance and reform in three consecutive phases, I search for its contribution to understanding the role of informational processes in dealing with environmental challenges. These two theoretical traditions are brought together in Chapter 4, in which the key idea of informational governance of the environment is formulated, elaborated on and debated.

Chapters 5 to 10 investigate different aspects of what I will from then onwards label informational governance. Chapter 5 especially looks into the developments in environmental monitoring and surveillance under conditions of the Information Age. How has environmental monitoring developed and what are the key innovations that contribute to informational governance? To what extent do enhanced monitoring and information collection possibilities result in larger surveillance, with all the teething troubles coming along with that? Special attention is paid to the globalisation of environmental monitoring and to the expanded role of citizen-consumers in counter-surveillance. Chapters 6, 7 and 8 focus on how respectively states, businesses and NGOs reorganise and redefine their role, position and strategies in environmental struggles and reforms, following the expanded practices and possibilities of informational governance. How do developments in information technologies, informational processes and information disclosures reconstitute the practices and institutions of environmental reform in which these actors are engaged? Chapter 9 specifically focuses on the major developments in the media and their role in informational governance of the environment. Are the media becoming the new major force and power in environmental reform, now that information becomes such a crucial resource and space in environmental controversies?

Although Chapters 5 to 9 focus especially on information-rich environments (OECD countries and the networks and infrastructures that make up the global network society), in Chapter 10 attention is turned to the informational peripheries, by focusing on two countries that are arguably not (yet) part of the informational centre: China and Vietnam. In assessing to what extent information governance has relevance in 'information-poor' environments, I investigate the actualities, forms and potentials of environmental informational dynamics in China and Vietnam, with respect to most of the themes addressed in the former chapters.

Finally, Chapter 11 draws conclusions on informational governance. Is it really emerging massively on the waves of an information age, replacing conventional modes of environmental governance? How do we tentatively judge these developments? And what is a research agenda for further investigating informational governance with respect to environmental challenges?

PART I

Theory

2 | *From Information Society to Information Age*

1. The transformation of modern society

In the 1960s and 1970s, the postindustrial society idea was developed by a growing number of respected authors, who started to investigate the discontinuities in modern society. These scholars believed and illustrated that modern society was moving from an industrial society to what they labelled a postindustrial society. According to them, this transformation meant a radical change of the modern order, comparable in magnitude to the transformation of the agricultural society into the industrial society. Publications of the famous Harvard sociologist Daniel Bell (1973, 1976, 1979), and more popular studies such as *Future Shock* (Alvin Toffler, 1970) and *The Age of Discontinuity* (Peter Drucker, 1969), started a period of intense and stimulating debate on the future of industrial society and the contours of a new phase of the modern order, markedly different from the heydays of industrialised modernity.

According to Kumar (1995: 2) the debate on the postindustrial society ended more or less with the oil crisis in 1973. From then onwards, the 'limits to growth', the containment of industrial capitalism rather than the dynamic potential of industrialism, dominated the debate. With deindustrialisation and economic decline as the main headlines in the newspapers, visions and agendas of a postindustrial society were no longer very attractive. But, at the same time, classic industrialism and industrialisation seemed no longer adequate to characterize Western society as it entered the last quarter of the twentieth century. There is an additional reason why the concept of a postindustrial society lost some of its attraction. Notions of *post* generally do not hold very long in labelling new times. They are basically referring backwards to characteristics that are no longer adequate (postindustrial, postmodern, postmaterialism and so on), without being clear on what has instead become the central common denominator of the new time.

It is especially under these conditions and circumstances that the notion of Information Society emerged. This notion should be seen strongly in line with much of the postindustrialist ideas. Daniel Bell, in his notion of postindustrialism, had already classified theoretical knowledge as the most important feature of the coming society. Later, Bell (1979) became an even more explicit supporter of ideas of the Information Society, by referring to the new information technologies invading every sector of society. Instead of postindustrialism, the new methods, processes and quantities of acquiring, handling and distributing information became the key feature of a new modern order in the making. Of course, debates and interpretation of the transformation of modernity did not end by launching the notion and ideas of the Information Society. As Kumar (1995) concludes, some social scientists took the idea of a turn, a radical transformation, of classical industrialism one step further, in interpreting the new social order in terms of postmodernity. Where the idea and conception of Information Society fit rather well in the Enlightenment character of Western thought, with an emphasis on rationality and progress and building on the main institutional traits of Western society, postmodernity consists of a transformation of these key Western notions and institutions beyond recognition. Although the notion of postmodernity raised significant support, criticism and debate for some time, also here the indication *post* was not there to stay.

In this chapter, we will especially explore ideas on the Information Society and how these ideas transformed into the Information Age literature. In that sense, we will make a rather selective review of four decades of literature that tried to interpret the transformation of modern society, first especially in the West but later more globally. The purpose of this selective review is to understand how various scholars have understood, interpreted and framed the role of information and information technologies in the transformation of modern society. It does of course not mean that I fully embrace all ideas and interpretations of Information Society and Information Age scholars, as will become clear throughout this volume. In focusing on these schools of thought, I will leave significant parts of the discourse on a changing modern order aside, for instance, that around postmodernism, only touching on it in clarifying Informational Society/Age scholars and their positions and literature.

2. The Information Society thesis

Several revolutionary developments in the late 1940s and early 1950s formed the foundation for the later claim that information and information technology had become the central resources and production forces, and the main ideology, structuring modern society. The birth of the computer, accomplished during World War II and in the direct after-war years in close relation with military requirements, stood central (Noble, 1986). As is the case for many scientific developments, such as those around nuclear energy and chemicals, military needs and military funding speeded up research and development in electrical circuits and computer technology (Saxby, 1990; see on the – modest – influence of the military on the development of Internet: McPhail, 2006: 291ff). Although "origins do not determine destinations" (Kumar, 1995: 7), the origins do teach us a vast amount about technological trajectories, motivations and major social forces shaping directions. And, as we will come to see later, these learning moments have been used especially by the critics of the Information Society.

As will become clear in this section, it is almost impossible to give one leading definition of the Information Society, as different authors emphasise different aspects and developments of what they see or interpret as key to the idea of Information Society. But, generally, it is the computer that stands symbol for and is put central in the development of the Information Society idea: "Computer technology is to the information age what mechanization was to the industrial revolution" (Naisbitt, 1984: 22). The computer transformed many operations of industrial society, but it has been especially the marriage of the computer with telecommunications that brought the Information Society into being, as an ideology and a material reality. It is from then onwards that not only scientists but also popular authors such as Toffler (1970, 1981) and Naisbitt (1984) started to include information technology in their analyses of a changing world order, of a third industrial revolution (Bell, 1987: 11). The combination of television, computers, telephone, microelectronics, satellites and fibre optic cable brought new flows and dynamics in information processing, making available all forms of information over ever-larger geographical stretches in decreasing amounts of time. But this was as much a quantitative change in information flows, as a qualitative transformation: "The Third Wave thus begins a new era – the age of the de-massified media. A new info-sphere is emerging alongside the new techno-sphere" (Toffler, 1981: 165). As many advocates of the Information Society noticed the time- and space-scapes of past societies were undergoing radical changes during what they saw as the emergence of the Information Society. The geographical space for social processes enlarged to the entire globe and the slow rhythms and tempo of nature were replaced by real-time, instant-time or timeless-time processes and interactions.

In his informative and rich overview of Information Society theories, Webster (2002) concludes that the analyses on the coming of the Information Society focused not only on indicators of the volume and direction of technological innovations, of the kind identified earlier. Webster identifies five sets of criteria or indicators (technological, economic, occupational, spatial, cultural) by which various scholars have tried to qualify, but also especially to quantify, the change from an industrial society into an information society. In arguing that these quantitative changes turned into a qualitative transformation, each of these sets of indicators have met their own problems of measurement, as we will see later. Questioning the measurement along these indicators is more that just a methodological problem; it touches the very essence of the question whether we indeed have radically and fundamentally changed from an industrial society to an information society.

So, the radical changes in the structural and material features of the Information Society come along with the rapid enlargement of the information sector in the economy. Machlup (1962), Bell (1973), Porat (1977), Stonier (1983), Castells (1996/1997), Wilson (2004) and others have reported on the rapid increase in several economic indicators when referring to the information sector in the late 1960s and 1970s. Jobs, gross national product (GNP) percentages and wages related to the informational sectors of society enlarged quickly in the 1970s. Using advanced and detailed calculation methods, Porat (1977), for instance, concluded that by the late 1970s 50 percent or more of the U.S. workforce was employed in the information economy and almost half of the U.S. GNP came from the informational sectors.[1] Daniel Bell

---

[1] Porat (1977) distinguished two informational sectors: the primary sector consisting of all direct informational organisations, and the secondary sector involving informational activities in other firms and state organizations (e.g., research and development sections of companies). Thirty years later, McPhail (2006) came to exactly the same figure: 50 percent of U.S. GNP is related to information-based services.

(1973), arguably one of the founding fathers of the Information Society idea, was especially interested in occupational changes: the decline of jobs in the manufacturing sector and their increase in the service sector, in which information workers were dominant. The work of Gouldner (1979) and Perkin (1989) on professionals can be read in the same tradition. These quantitative calculations and figures have been questioned and debated by others. The Information Society adherents have been accused of using data that is too aggregated to distinguish different kinds of informational activities and jobs. Do all employees in the service sectors or even in the informational organisations really perform informational work? But it is not just quantitative economic indicators that have taken a central position in the economic arguments on the Information Society. Charles Leadbeater (1999), for instance, especially looked into what this occupational change and the increasing importance of information means in qualitative terms. Knowledge, skills, talent, creativity and the capacity to exploit networks have become the bases of obtaining wealth, instead of physical effort. This would even require the reformulation of economic theory itself. Labour and capital, considered to be the central variables and forces of production, are to be complemented, or even replaced, by information and knowledge; the labour theory of value by the knowledge theory of value (cf. Kumar, 1995: 12).

Although dominant, the assessment of the coming of an information society has not remained restricted to changes in the economic-technological domain. Bell remains cautious in claiming an overall transformation of all sectors of society, as he keeps to the idea that different domains or realms in society are structured and governed by different dynamics, factors, principles, rationalities and rhythms of change. And the informational change, according to Bell, is mainly related to the technoeconomic structure of modern society, leaving political and cultural domains rather intact. Others, however, are less cautious and follow the basic logic of revolutionary change, which claims interrelated transformations of all sectors and domains of a society. This is how sociologists and historians have understood the Industrial Revolution. Toffler (1981) relates the coming of the Information Society to changes in not just the techno-sphere, but also in what he calls the socio-sphere, the power-sphere, the bio-sphere and the psycho-sphere. The Information Society is then not just a new mode of production, part of a transformation of the economic structure and infrastructure, but a complete restructuring of all institutions of modern society, including political and cultural sectors.

The transformation towards an Information Society is then believed to have as its basis new technologies, new forces of production. But these technologies influence and – according to some – determine new forms of life and living in various sectors: work, leisure, at home, family relationships, politics, culture and so on. Overall, Information Society theorists tend to apply an optimistic, positive, evolutionary approach towards such transitions, strongly rooted in the modernist and rationalist traditions. New democracy, egalitarianism, a classless society, decentralized and horizontal institutions are some of the changes believed to follow from the centrality of the computer. Thus, Information Society analysts often – but not always, as we will see later – turn into Information Society propagandists and advocates.

The Information Society and the environment

The emergence of the literature and debate on the Information Society paralleled in time what I have labelled elsewhere the second wave of environmental concern (Mol, 1995). From the late 1960s untill at least the mid-1970s, Western industrialized societies were not only confronted with increased attention for environmental problems and natural resource depletion by large segments of civil society. These societies also witnessed major institutional innovations with respect to the environment: new political institutions such as ministries for the environment, legal frameworks and environmental impact assessment systems; new NGOs such as the World Wildlife Fund (WWF) and Friends of the Earth (FoE) and mobilization campaigns such as Earth Day; new periodicals, media specialists and educational institutes and new research, innovation and education institutions that focused on the environment (see Chapter 3).

Given these time parallels, it is remarkable how little of these environmental interests and considerations have been at the centre of the Information Society ideas and literature. In the overviews of Kumar (1995) on the Information Society literature, of Badham (1984; 1986) on the postindustrial society debate, and, more recently, of Webster (2002) on theories of the Information Society, the environment hardly plays any role.[2] Although the general conclusion should indeed be that

[2] In an edited volume (Webster, 2001b) on how in the Information Age information and information technologies are transforming politics and culture,

there is hardly any reflection on environmental crisis and reform in the postindustrial and Information Society studies, there are some notable exceptions. In the literature on the Information Society in the 1970s and 1980s, environmental considerations and arguments prevail – be it marginally – in four different ways.

The first theme in which the Information Society ideas link to the environment is via the postmaterialism thesis that has especially become known via the work of Inglehart (1977; 1990). Although Inglehart himself hardly related postmaterialist values (freedom, self-expression, aesthetics and quality of life) to the idea of the coming of the Information Society, several other scholars have drawn loose parallels – often in the form of correlations and sometimes in terms of causal relations – between postmaterialism and the Information Society.[3] The Information Society idea, with its emphasis on service sector jobs, informational processes and nonphysical and nonmaterial capital as the basis for development, would fit very well the emergence of post-materialist values, leading away from the strong material basis of industrial society, its blue-collar labour force and its conventional hierarchic forms of governance. Bell's (1973) 'new consciousness' resembles the postmaterialist values put forward by Inglehart. But, at the same time, Inglehart (1990) prefers to speak of advanced industrial rather than postindustrial society, distancing himself from the postindustrial and Information Society scholars.

Second, the emphasis of some of the Information Society literature on decentralization and democratisation links strongly with what stood central in the 1970s and 1980s in both the environmental social science literature and the ideas of environmental movements. *A Blueprint for Survival* (Ecologist, 1972) and the work of Murray Bookchin (1982) and Schumacher (1973) inspired many environmental activists and organisations (such as the well-known Mondragon cooperatives in Spain) towards more decentralized forms of production-consumption relations and decentralized modes of governance in especially Europe and the United States. The quest by especially the more left-wing

the environment is much more present, especially in the form of the environmental movement and not so much in wider categories and institutions of environmental governance (such as those related to politics or economic processes).

[3] For the moment, I put between brackets the interesting debates of whether postmaterialism indeed relates to environmental interests and values. See, for balanced critiques on such positions, Neil Carter (2000: 89–91) and Michael Bell (2004: 161–162).

environmentalists for decentralization parallels the ideas of Information Society scholars such as Daniel Bell, without any noticeable direct influence between the two.

Third, the Information Society thesis overlaps with the environmental agenda of the 1970s in emphasizing deindustrialisation. Where the Information Society literature investigates, hypothesizes and advocates the switch form the industrial sector to the service sectors, environmentalists in the 1970s joined them in calling for a deindustrialisation of modern society, as the industrial sector stands for massive emissions and abundant and inefficient (mis)use of natural resources. This parallel is most strongly felt in Boris Frankel's (1987) analysis of what he labels the Postindustrial Utopians, drawing parallels between Information Society theorists such as Toffler and Bell and environmental scholars such as Rudolf Bahro and André Gorz. The deindustrialisation agendas of both groups of authors show remarkable commonalties.

Finally, the dematerialization ideas of the environmental movement and scholars often find some rationale in the hypothesis of the coming of the Information Society, especially as in both notions the material basis of production and consumption and of economic development is increasingly replaced by nonmaterial forms of production and consumption. Again, the growth of the service sectors are part of this, but also the fine-tuning of industrial production processes with the help of ICT, the decrease in travelling as a result of improved communication systems and ideas on the switch from material products (newspapers, magazines, books) to digital alternatives. Even virtual tourism is stressed in both traditions.

Still, in all, it remains rather remarkable how absent environmental arguments and perspectives have been in the debates on the Information Society in the 1970s and 1980s. There is, for instance, hardly any reference to the possibilities of environmental movements to use ICT in their campaigns; there is no emphasis on improved environmental monitoring and information collection possibilities following the Information Society and there is no attention paid to the environmental improvements of an information-based economy. Clearly, both schools of thought come from very different traditions, hardly sharing common intellectual and societal backgrounds and advocates. It was only in the 1980s, when environmental scholars and advocates started to develop a less hostile attitude to technological innovations, that this started to change. Joseph Huber (1982) was perhaps one of the first to fully acknowledge the significance of the information revolution for environmental reforms, as will be further elaborated in the next chapter.

Assessing and debating the Information Society claims

In the debate and discussions following the Information Society thesis, there exists remarkable agreement on the significant impact information and information technology have on various aspects of life. Even the most severe critics of the Information Society thesis join in noting the massive consequences of the information revolution on, among others, households and homes (Miles, 1988), on lifestyles, on work and on the capitalist economy (Morris-Suzuki, 1984). Information technology is generally not seen as just another new technology but identified as a revolutionary technology, causing radical changes.

But the acceptance of the fundamental contribution of information technologies and information to modern society does not automatically result in acceptance of the Information Society thesis in its full breadth: the thesis that a new society has emerged, fundamentally different from the 'old' one. Not surprisingly, criticism against the idea of an Information Society parallels to a large extent the objections against the postindustrialism thesis, as both theses are so strongly related to each other. We can summarize the objections against the Information Society thesis in five main points.

First, there is significant criticism on the statistical and quantitative approach in claiming the coming of the Information Society. As noted earlier, much of the Information Society literature qualifies the changes taking place along various indicators, claiming that massive quantitative changes (in jobs, GNP composition, technology generation and penetration, etc.) result in qualitative transformations of society at large. Roszak (1986) and Webster (2002: 21–23), among others, denounce the 'more-quantity-of-information-to-new-quality-of-society' argument. According to them, Information Society scholars put together all kinds of different sorts and forms of information (or information workers, informational GNP, information sectors), disguising the fundamental differences in (sorts, forms, dimensions of) information and their impacts on the wider society. Without looking into the quality of information, the economic activities related to information, or the specific occupational patterns, data remain far too general and aggregate to draw any lasting qualitative conclusions.

Second, the impact of information technology on employment and on the skills and autonomy of workers has been exaggerated, as we can conclude now. Although information technology, as every significant new technology, has shifted employment within and between sectors, these shifts of employment in and between the industrial production and service sectors has not resulted in structural higher levels of unemployment, or significant and structural lower working hours per week (such as was especially argued among the left-wing propagandists of postindustrialism or the Information Society; cf. Frankel, 1987). Equally, the Information Society ideas on shifts in the quality of work towards increased skills, large autonomy and more professionalisation have been questioned. Although the number of professional, administrative and managerial employees has increased sharply, this does not lead us to easy overall conclusions on higher skills, larger autonomy and more professionalisation. Similarly, several critics concluded in the 1980s that employment growth has not been based so much on the advanced knowledge sector but came, rather, from the lower levels of the tertiary sector (eating and drinking, health sector, routine information workers such as data processors, etc.).

There is a strong role of the state and government, and especially of the military sectors of the state, in research and development for the information technology and information sectors. Large private multinationals, such as IBM, AT&T, IT&T, Philips and Siemens, link with these state and governmental sectors in the further development of information technology. Increasingly, these firms aim to become integrated information concerns, rather than having a stronghold in just one of the information sectors. Information technology had clearly become big business in the 1970s, and its prime goals have been instrumental to capital, according to Schiller (1985). Third, this has clearly undermined the whole idea of the Information Society as a process of democratisation, and a means to create a new, progressive society with more prosperity. According to critics such as Morris-Suzuki (1988) and Lyon (1988), information technology was, first and foremost, developed in the interest of capital, the administrative elites and the military establishment. "Capitalist industrialism has not been transcended, but simply extended, deepened and perfected" (Walker, 1985: 72). It is only the rich nations, the elites and specific economic sectors that have profited from the information revolution, which makes the claim of a new and better society awkward, according to these critics.

Fourth, according to its critics, the Information Society thesis falls short in its short-sighted historical perspective and as such neglects long-term developments. From different perspectives and with various empirical and theoretical evidence, a number of authors have argued that the changes emphasized in the Information Society thesis are rather part of long-term historical developments (and not of sudden revolutionary changes in the third quarter of the twentieth century). Several economists have interpreted the Information Society developments as not much more than a further application of Taylorism (cf. Braverman, 1974), or part of a much wider and longer-term Control Revolution (Beniger, 1985). Work under conditions of an Information Society is as much part of routinisation, fragmentation, deskilling and subject to 'scientific management', as blue-collar work was during the heyday of industrial society (and following the Industrial Revolution). Computers have hardly changed fundamentally the organization and control of economic processes.

Fifth, and last, the idea of the Information Society is an ideology propagating radical transformation, whereas in fact information technology only brings fundamental continuity. Existing patterns continue rather than showing discontinuities: work and leisure are further industrialized, social inequalities do not diminish, political arrangements and institutions continue and capitalist relations of production are far from abandoned but, rather, refined. When the importance of information, knowledge and information technology increases in society, the main consequences are not cultivating wisdom and free access. Rather, with the Information Society knowledge and information become privatised, commodified, appropriated for marketising and profit. Information becomes a mass-marketed product in and of itself, not unlike what befell all previous new services and technologies. Information technology has been shaped, developed and used in conformity with the dominant social and political interests. And although these interests cannot determine all details of information technology structures, outcomes and effects, they push for continuity rather than for radical change. Quite recently, Paehlke (2003) has repeated this criticism, in stating that the central axial principle of what he calls electronic capitalism is capitalist growth, and not the codification of theoretical knowledge, as Bell claims.

The backside of the Information Society

Although most of the studies on the Information Society thesis have rather optimistic and positive connotations when analysing the transformations in technological paradigm, occupational patterns, economic structure and social and cultural institutions, there is also a vast literature that is more critical and ambivalent towards the coming of an Information Society. In contrast to some of the sceptics mentioned earlier, these critics do not so much deny the (coming or prevailing) transformation into an Information Society. They concentrate, rather, on the negative effects and structural design faults of such a new society in the making, highlighting especially three drawbacks.

First, and related to the argument of continuity of a capitalist structure under conditions of an Information Society, several authors have strongly criticised the concentration of information flows in the hands of a few powerful multinationals and elites. The critical theorist Herbert Schiller is arguably among the strongest critics of a positive connotation of the information revolution. For whose benefit and under whose control will the Information Society be implemented? Schiller (1969; 1981; 1989) basically provides three interrelated arguments in emphasizing the inequalities and the power relations related to the Information Society. First, information and communication innovations are developed and implemented under market conditions and, thus, decisively influenced by market pressures, inequalities and profits. The commodification of information means that information is treated like any other good or commodity under capitalism. Hence, these information commodities will benefit few, whereas large sections of society will be deprived of these benefits. Second, the capacity to generate information, the distribution of it and the access to information are significantly determined by and run along class inequalities. Consequently, unequal class positions determine the winners and the losers of the informational revolution, in as much as it did with the Industrial Revolution. Third, and finally, the dominance of contemporary corporate capitalism, formed by oligopolistic, concentrated transnational firms, means that the Information Society will resemble these hegemonic interests. These three characteristics prevent Schiller from any optimism on the outcomes of an Information Society. As capitalism continues also in the Information Society, there is nothing really new to report in terms of inequalities and dominations.

A second, related argument has especially been developed within UNESCO and the call for a so-called New World Information and Communication Order, NWICO.[4] Through the concentration of media and information in the hands of a limited number of states and corporations of the North, the Information Society will result in an unbalanced (in terms of quantity and substance) flow of information from the North to the South, heavily controlled and biased by the cultural and hegemonic interests of northern actors. After the stages of military colonialism, Christian colonialism and mercantile colonialism, we are since the 1950s in a phase of electronic colonialism (McPhail, 2006). There is a strong cultural and informational dependency of poorer regions of the world on the postindustrial nations and (their) multinationals, established by the dominance of the latter two in the information and communication hardware and software. This situation has only worsened with the neoliberal developments and privatization of the media, as states are increasingly out of control, or can legitimately claim to be out of control. Hence, the NWICO adherents (many of them coming from developing countries) argued for a restructuring of the information order, providing for a stronger dominance and control of nation-states (especially those of the South) on the production and flow of information, and a limitation of the freedom of transnational information flows.

The third main negative (side) effect of the Information Society is related to the enhanced possibilities for states, but also for powerful economic actors, to control and monitor civilians in their daily life and practices. Following the work of Michel Foucault, the theme of surveillance is more than incidentally linked to the coming of the Information Society (cf. Champbell and Connor, 1986). Whether related to the military and the police, to 'transactional information' (Burnham, 1983) of purchases and money transfers or to monitoring daily practices of citizens, the systematic collection, processing and storage of information is thought to be problematic, especially because in contemporary societies this relates so strongly to unequal power relations, discipline and loss of privacy. Surveillance through the use of computers has been related to the nation-state, to bureaucracies (in the Weberian sense), to technologies (in line with Ellul's work) and to the political economy (where it is related to conflicting interests and economic power

---

[4] See Chapter 9 for a further analysis of the NWICO debate.

relations, for instance following the work of Braverman). David Lyon (1988; 2001) has perhaps been mostly associated with what is now called 'surveillance studies' or the 'surveillance society'. According to him, computerisation and digitalisation have enabled surveillance far beyond what was originally envisioned by Foucault and his contemporaries, leading to a superpanopticon and hypersurveillance.

3. The Information Age

Although academic persuasion of, research into, and debate on the Information Society continued in the 1980s, the momentum, innovation and quantity of this research line and school clearly stagnated. Although a significant group of scholars remained attracted to ideas of the Information Society, in general the academic social sciences as well as the general public shifted to other schools and ideas in the 1980s. It is only in the 1990s that ideas on the Information Society, on the central importance of information and on information technology received a new impetus, be it framed and conceptualised partly in a different way. And a different group of authors became the leading spokespersons, among them Manuel Castells, John Urry, Anthony Giddens, Saskia Sassen, Ulrich Beck and Scott Lash. Not all give information and information technology an equally important place in the transformation of modernity, but they share a common frame of reference, in which globalisation is closely related to information (flows and infrastructures), networks and uncertainties.

In his impressive tetralogy on the history of industrial societies from 1789 to 1991, the historian Eric Hobsbawm identifies four periods, each defined by a characteristic *leitmotiv*: the *Age of Revolution* (1789–1848), the *Age of Capital* (1848–1875), the *Age of Empire* (1875–1914), and the *Age of Extremes* (1914–1991). The fundamental concepts express the 'structuring principles' of those eras, perhaps less perceived by their contemporaries than by historians studying social developments in these time periods. To understand developments in different social sectors of industrial societies of that time, the analytic perspective should start from such a fundamental *leitmotiv*, as Hobsbawm argues. Manuel Castells's major trilogy on *The Information Age* should be understood in this tradition, claiming that informational developments are the key *leitmotiv* to understand the present discontinuities in the global order. I will use Castells's label Information Age to contrast the ideas and framing in this period to the Information Society literature of the 1960s and 1970s.[5]

After briefly introducing Castells's major work to better understand the context of this second group of authors, I will compare the two periods in how they frame informational developments. I will do so by first analysing the differences – focusing on three main aspects: globalisation, uncertainty and governance – to finalize with the main similarities and continuities between the Information Society ideas and the literature on Information Age.

Castells's network society

In his three-volume opus magnus, Castells (1996/1997) analyses in rich empirical and theoretical detail how in the 1980s and 1990s a new social morphology emerged through globalisation and the constitution of a new technological paradigm (called informationalism), coining this transition as the coming of the global network society. The global network society is a new way of structuring time and space through reintegrating the functional unity of different elements at distant locations made possible by modern transport, information and communication technology (Castells, 1996). Networks are the key constituting units, both in terms of physical infrastructures and social systems. The core activities that shape and control human life around the globe are organized in networks. As indicated earlier, Castells is not the first to analyse the role of information and networks in reconstructing the modern order. But his analysis is radical in that he replaces conventional

---

[5] These labels should not be taken too absolutely: also in the 1990s and in the new millennium the term *Information Society* is still regularly used (e.g., by the European Commission in launching their research program on Information Society Technologies; in the 2005 World Summit on the Information Society), and in the 1980s incidentally authors referred to the Information Age (cf. Naisbitt, 1984). It is partly for pragmatic reasons that I use these labels, although they generally do refer to different substantive writings. The notion of 'age' seems more adequate than 'society' for the second group of scholars, as in the new millennium several of them consider society a zombie-concept (e.g., Beck and Willms, 2004; Urry, 2003; Castells, 1996/1997). See also Castells (2004: 6–7) for counterpositioning the concept of Information Society and his own work on the Information Age.

concepts and controversies of the social sciences with new ones. States, societies and physical space are no longer the core concepts for understanding modern societies in the Information Age, according to Castells. Flows and networks replace them. Central in his analysis is the tension between the 'space of flows' and the 'space of places'.

The new institutional makeup of the network society is to be understood in direct relation to a new layer or dimension emerging within and in between our societies. This new layer, the 'space of flows', should be understood not as a new layer in the geographical meaning of the word, but, rather, in terms of a new kind of time-space organization of social practices. The space of flows refers to new social dynamics – to new concepts of time, space and power. But the notion of space of flows also has a substantial connotation in Castells's work. The dominance of the space of flows then refers to the power elites that operate at the most crucial nodes of the global networks, knowing best how to handle the switches, codes, and programs (cf. Castells, 2004), which govern global flows of money, capital and information, at the expense of the vast majority of ordinary people living their lives in the space of place.[6] The only option left to the locals is protest and resistance against the disturbing and exploitative character of the space of flows. Although Castells is the first to argue that the logic of the space of flows will not be displayed in the network society without resistance, his analysis of the new constitution definitely has a deterministic ring to it, with the space of flows performing as a stand-in for a very powerful class of global capitalists. From this perspective, the new social order of the network society should not be seen in association with the positive image of the new dynamics of 'the Internet society' (as several predecessors of the Information Society did). Instead, this new social order manifests itself as a *meta-social disorder* (Castells, 1996: 477), an order derived from the exploitative and uncontrollable logics of markets, genes and (information) technology.

In that sense, Castells's network society analysis combines elements of both the optimists of the Information Society as well as their critics. But Castells's ideas, as well as those of his contemporaries in this

---

[6] With the space of places, Castells (1996: 378, 423ff) refers to the place-based spatial organization of social life, as commonly perceived and experienced by the majority of citizens in advanced and developing societies.

tradition, differ on at least three points fundamentally from the analyses of both adherents and critics of the Information Society.

Globalisation

If anything distinguishes the literature on the Information Society of the 1960s and 1970s from the theories on the Information Age of the 1990s and beyond, than it is the centrality of globalisation processes. When the first commercial satellite began circling the earth by the end of the 1960s, enabling and enhancing global communications, we entered a new phase of modernity. And many Information Society scholars in the 1970s pointed already to the potential these new developments bring to global communication systems. But the full contours of the ways in which these global communication and information systems interfered with, strengthened and coproduced other globalisation dynamics became only visible in the 1990s. It was only by the end of the 1980s that the concept of globalisation became widespread, to dominate the social and political science literature and debate of the 1990s. Elsewhere (Mol, 2001), I have detailed how the discontinuist school[7] in globalisation studies especially has argued that globalisation has recently fundamentally altered the modern order.

Information, information and communication processes and information and communication technology form central elements in much of the discontinuist globalisation interpretations and literature. Anthony Giddens (1990), for instance, emphasizes the significance of the new telecommunication and information technologies in accelerating the compression of time and space and thus contributing to a qualitative change in globalisation. These technologies and the related organizational innovations have altered the scope and speed of economic decision making. This enhances the capacity of the economic

---

[7] The discontinuist school of thought contrasts with the continuity scholars, who argue that globalisation is not really something new but dates back as far as the sixteenth century: "The widespread view that the present degree of globalization is in some way new and unprecedented is, therefore, false" (Glyn and Sutcliffe, 1992: 91). Discontinuist scholars such as Anthony Giddens, Phil McMichael, David Held, and Manuel Castells, in contrast, argue that the forms and dynamics of global interconnectedness and interdependence have changed fundamentally in the last thirty years, bringing about a new social order.

system to respond rapidly to fluctuations, but it also renders it more vulnerable to overreaction to relatively minor disturbances, thereby producing major crises. Manuel Castells is even more outspoken on the role of information and communication technology in bringing about the global network society. Globalisation is directly related to informational processes and technologies in understanding the emergence of a new social morphology, the global network society. It is the centrality of informational processes in combination with globalisation that make us have to use new concepts to understand the global network society. Building on Castells, John Urry (2003; 2004) is arguably the most radical scholar in doing away with 'zombie' concepts such as state and society under conditions of globalisation, putting (information) flows and networks as the new architectures of a global modernity.

The key position of globalisation processes in assessing the transformations taking place in modernity is reflected in the core concepts that are used. Where the Information *Society* literature focused on the changes in the national economies and societies, the Information *Age* scholars were and are particularly concerned with how global networks and flows of information (and capital and persons) restructure the world. Late modernity, reflexive modernity, second modernity and global modernity all refer to these global changes and transformations, much more than notions of postindustrial society, Information Society, home-centered society or – an arbitrary case – the Risk Society.[8] Consequently, the argument of a fundamentally different social order through the information revolution is much more powerful in the Information Age literature, as it strongly connects to globalisation arguments. Current scholars in global communication and information studies follow that line of argument by doing away with first modernity concepts and notions. McPhail (2006: 331), for instance, claims that we are presently in a postsovereignty reality, in which the Internet and cyberspace make that 'industrial-era concepts such as space, location, control, bricks and mortar, and monopoly are marginalized.'

---

[8] Ulrich Beck in his 1986 *Risk Society* only indirectly emphasises the *increasing* role and influence of information processes on the making of a new social order. This did change in his more recent work on reflexive modernity, globalisation and the World Risk Society (cf. Beck et al. 1994; Beck, 1997; Beck, 2004; Beck and Willms, 2004), although the core of his analysis remained the same.

Although not all scholars in the Information Age tradition put as much weight as Castells on the contribution or centrality of informational developments to the changing constitution of the modern global order,[9] most of them acknowledge that information flows and the new technological paradigm (based on the Internet and ICT) are at the basis of fundamentally different global economic, political and social processes.

Knowledge and uncertainty

In his analysis of the emergence of a *Risk Society,* Ulrich Beck (1986) forcefully put radical uncertainty on the agenda of late modern societies. This radical uncertainty is especially, although not only, linked to dealing with environmental and food safety problems. Although initially Beck draws heavily on a select number of examples (Chernobyl, mad cow disease or BSE, climate change, genetically modified organisms [GMOs]), his later analyses on structural uncertainty move partly beyond such a limited set of (environmental) issues, claiming that these uncertainties are related to the overall structural characteristics of late modern society, and not so much to a particular set of (new) environmental challenges.[10] As conventional science has lost its Enlightenment character as well as its authority with respect to monitoring, measuring, interpreting and making truth-claims on environment and health risks, and no other authority has stepped in, late modern society is faced with an almost inherent uncertainty with respect to these (and other) dangers and risks. The widely available and constantly growing global flows of information and contrainformation on (potential) risks and threats form part of the causes of science losing

---

[9] Compare, for instance, the centrality of information and information technology in the work of Manuel Castells (1996/1997; 2001) with the more modest place of information in the studies of Anthony Giddens (1990; 1994), John Urry (2000; 2003; 2004) or Ulrich Beck (1994; 1996; 1997; Beck and Willms, 2004).

[10] But Beck remains ambivalent as to the definition of environmental and food safety problems that can be understood with the notion of the Risk Society. The expropriation of the senses, the 'Fahrstuhl-effect' and the lack of trust are all developed with respect to a specific new category of risks, for which the old institutions of high or simple modernity seem no longer relevant. The question is whether these insights have the same relevance for, for example, domestic solid waste, industrial water pollution and eutrophication caused by farming.

its monopoly and authority, and contribute to growing feelings of uncertainty among large segments of the population. For all kinds of risks, dangers and (environmental) threats experts are challenged by contraexperts. With no scientific and undisputed mechanism – no litmus test – for closing these controversies and debates, lay actors (but also decision makers) are left in uncertainty about the actual risks of eating meat, organic food or GMO food. Being able and capable to deal routinely with both structurally contested information and uncertainties seems to have become one of the necessary competencies of late-modern citizen-consumers. In this tradition, a whole new school of thought emerged, with, for instance, authors such as Brian Wynne, John Hannigan and Steven Yearley. Recently, Matthias Gross (2006) added to this literature by analyzing how new knowledge and inventions always create new ignorance, uncertainties and what he labels nonknowledge.

Following his work on the Risk Society, Beck's project of reflexive modernity aims to analyse the ways in which modern society deals with these informational and knowledge uncertainties, both individually and institutionally, by constructing reflexive practices and institutions that build trust, set limits to our doubting and arrive at answers and conclusions in order to be able to get things done and move on (cf. Beck and Willms, 2004). Here we see the contours of a more constructive interpretation of growing information flows and related uncertainties, in which new knowledge and information constantly questions and challenges existing patterns, structures and knowledges, without necessarily leading to apocalyptic prophecies and stalemate positions as a result of fundamental and ongoing uncertainties.

John Urry (2000, 2003) further explores and radicalises the notion of uncertainty in his sociology of networks and flows. Although he judges Castells's trilogy on the rise of the network society as the best effort so far to analyse networked modernity, Urry sets himself the task of elaborating and refining the conceptual apparatus as used by Castells. Instead of Castells's dichotomy of 'space of place' and 'space of flows', Urry suggests approaching spatial patterns in three ways or modalities, distinguishing among regions (i.e., objects geographically clustered together), globally integrated networks (more or less stable, enduring, and predictable relations between nodes or hubs, stretching across different regions, with relatively walled routes for flows) and, finally, global fluids (spatial patterns determined neither by boundaries nor by more or less stable relations, but by large flexibility and liquidity). Where Beck basically relates (informational) uncertainty to contested science and knowledge and Castells does hardly problematise knowledge or information in terms of being fundamentally uncertain or contested,[11] Urry puts uncertainty in a much broader perspective of global fluids. Uncertainty is not only related to measurements, information, facts and risks, but has become a structural property of global flows/fluids that constitute a global complexity. A global complexity that is inherently unpredictable, unmanageable, unexpected, chaotic, constantly on the move, unknown and thus fundamentally uncertain.[12] As such uncertainties relate to the outcomes and effects of all kind of social (and material) networks and flows, and – compared to Castells – information flows and ICT are considered less fundamental to the coming of a new – fundamentally uncertain – social order. In that sense, it becomes logical that Urry (2004) recently acknowledged the significance and influence of Ulrich Beck in understanding global complexity.

Having worked with Urry at various moments (e.g., Lash and Urry, 1987, 1994), Scott Lash is not too far from Urry's analysis. But Lash (2002) focuses much more strongly on information itself (and is in that sense closer to Castells), and – compared to Urry – takes a more cultural turn. According to Lash, contemporary times should be understood as an Information Age – and not as postmodernity or the Risk Society – in that the historical transition is one of a national, organized manufacturing capitalism to a global informational capitalism based on the spaceless mobility of idea-capital. Every sphere of life becomes increasingly mediatised and digitalised through information and disinformation. Much in line with the work and ideas of Marshall McLuhman and Jeremy Rifkin (2000), Lash's new information order replaces ownership and physical property relations by relations of access and intellectual capital. Power, inequality and wealth are reconfigured around

---

[11] Castells is more focused on the informational processes, rather than on the content or substance of information. Uncertainty in Castells's analysis is thus not related to science and scientific knowledge but to uncertainty of economic producers in a globalised network economy, in which they lose sight and control on what is happening in their social and economic environment (cf. Castells, 1996: 153, 190, 193).

[12] From a different analysis, Wallerstein reaches the same conclusion in claiming that the mutation of the World-System has now arrived "in the true realm of uncertainty" (Wallerstein, 1991: 15).

information-driven cultural systems, access/exclusion to information and control and domination over information technologies, networks and flows. The consequences are far reaching. The Information Age is at the same time a disinformation age, in which information is out of control through overloads, misinformation and disinformation. At the same time, digitalised information has immediacy; its working, effects and power are immediate and global, without any time for reflection.[13] Similar to Urry, Lash identifies the Information Age as one with structural unintended consequences, leading to disorder but also to new ordering principles and institutions, such as intellectual property rights, branding, Google and weblogs. These new institutions reorder the Information Age although along very different lines, institutions and mechanism compared to the former industrial society.

This is all (fundamentally) different from the Information Society literature and ideas, in which information and knowledge as such are not questioned. Information and knowledge are seen as unproblematic categories, still strongly in the Enlightenment tradition of emancipation. If anything, it is the unequal access to and monopolistic production/transmission of information that receives attention in the Information Society literature, but not the fundamental uncertainties, immediacies and disinformation that come along with global digitalized information flows. Structural uncertainties, the disenchantment with science, the undermining of conventional institutions and the constitution of new alternative decision-making practices and ordering institutions are typical products of, and only emerge in, the Information Age literature of the 1990s.

Governance

With globalisation and fundamental uncertainties of knowledge and information, ideas of governance are also different. Although Castells rooted the informational transformations and the rise of the network society primarily in the economy (as most Information Society scholars did), he also noted significant origins in and consequences for culture,

---

[13] The central argument of Lash is that in the Information Age with its dominant immediacy of information, (academic) reflection and critique becomes radically different, if not impossible. Although interesting, this is beyond the scope of argument of this book and will not be explored further. See Sandywell (2003) for a further analysis and criticism of Lash's position on this.

politics and governance. Politics and governance in the global net-work society are in some respects fundamentally different from con-ventional politics. From Castells's work, we can mention five points that are of specific relevance for our analysis. First, states lose much of the political and governance primacy and power in the global net-work society, an idea that is radicalised in Urry's (2003) recent work on Global Complexities. Second, politics in the Information Age is characterized by the centrality of informational networks and media. "Outside the media sphere there is only political marginality" (Castells, 1996: 312). Activists, politicians and firms adapt their politics to the media: symbols, sound bites and media presence become key issues. This also means that the political battleground moves to information, symbols and the media. Third, the players in informational politics and governance no longer have any possibility of effective control, as the networks are simply too fluid, too leaky and too undisciplined to control final political outcomes. Fourth, there is a primary importance of social movements in informational politics, as they offer collective identities, are symbol mobilisers and are champions in new forms of political agitation, connecting transnational networks with particular locations. Fifth, Castells sees possibilities for revitalizing and empow-ering localities by electronic connections and participation to political debates; hence, a (yet to be proven; cf. Webster, 2001b) potential for informational or electronic democracy.

There seems to be wide consensus among the Information Age schol-ars that a focus on state-based forms of national government is no longer very helpful in understanding today's complexities – in terms of diversity of actors, institutional arrangements and levels – in steering and governance. Just as science loses its monopoly on truth-claims and is faced with legitimacy problems, the nation-state also faces problems of trust, legitimacy and sovereignty under conditions of information-alism and global complexities. A whole new vocabulary emerges to pin down the innovations in governance arrangements, which can be witnessed as an answer to the limitations of nation-state governments. These new concepts converge in the idea that the nation-state loses its monopoly, but diverge in pinning down what complements or replaces the nation-state. Subpolitics, hybrid arrangements, multilevel gov-ernance, network governance, nonstate market-based arrangements, governance without government and political modernisation are just a few of these new conceptualisations.

In that sense, the Information Age ideas depart from the state centeredness of the Information Society literature. And this departure is closely linked to not only processes of globalisation, but also of the growing importance of a new technological paradigm, with ICT, cyberspace and information flows. The emergence of electronic newspapers, interactive cable, the Internet and direct satellite broadcasting indeed raises questions about the role of state regulations and the concept of national borders and boundaries. Kobrin (2001) is illustrative of a number of authors who point at the unique sets of governance questions that are emerging around cyberspace, the Internet and global information flows; questions on deterritorialisation, complexities, the irrelevance of space, the postsovereignty era and others.

4. Continuities between Information Society and Information Age

Notwithstanding the significant differences between the Information Society literature and the Information Age scholars, there are clear lines of continuity that justify discussing both in one chapter. Let me give four illustrations of the major continuities, beyond the general idea of the centrality of information (technology) in the constitution of a new modern order.

The continuities between the Information Society literature and the Information Age theorists are perhaps best stressed by Kumar's (1995) major study on new theories of the contemporary world. He starts his analysis with the postindustrialism/Information Society literature of Daniel Bell and his contemporaries and ends with interpreting the globalisation literature as the last version of similar studies that try to define and interpret the changing constitution of modernity. In that sense, according to Kumar, these studies and analyses all belong to the same family: interpretations of the (new) constitution of the modern order.

Second, much of the concrete empirical research that was carried out in the 1970s to assess whether a change in quantities (of the economy, the occupational patterns, of the information technologies, of speed and volume of transactions) resulted in a qualitative different phase of modernity is still done today. Today's studies on occupational patterns and their shift, on the contribution of informational sectors to the national economy of GNP (cf. Wilenius, 2002), on the diffusion of information technologies and on processing capacity and speed are not fundamentally different from those of Bell (1973), Porat (1977), and their contemporaries.

Third, it is also remarkable how close Castells and Bell come when they both investigate the main drivers for their new society: these drivers are to be found in the economy and in new technologies, whereas politics and culture take a second place. If one reads the critiques on *The Information Age,* the similarities with the comments issued against the Information Society thesis of Daniel Bell and his contemporaries is again striking. Garnham (1998) and Webster (2001b; 2002) have collected a number of objections against Castells's overly technological determinism, his reported empirical developments (and their measurement) towards an informational order and his emphasis on the transformation/innovation of his new order (while neglecting continuities). This criticism sounds all too familiar after the Information Society debate in the 1970s and 1980s (as has been mentioned earlier). In a similar way, Paehlke's (2003) electronic capitalism thesis is quite similar to Schiller's (1969, 1981) work on the marketisation, inequalities and monopolies that come along with the Information Society.

And, last but not least, as in the Information Society literature, also in the Information Age studies environment is only marginally emphasized, and then in a very specific way. If the environment comes to the fore in the Information Age literature, it is to assess the direct environmental consequences of a further informatisation, digitalisation and computerisation of society. The paperless office, teleworking and e-business are still the milestones to review the Information Society and Information Age on its environmental performance. Equal to the Information Society studies, only incidentally the Information Age literature relates information systems, networks and flows to a globalised environmental movement (e.g., Castells, 1997; 2004), to environmental governance and to environmental information systems related to economic activities that criss-cross borders. It is this omission that this volume aims to repair.

5. Conclusion

At various times in the past four decades, scholars have pointed at advances in information, informational processes and information and communication technologies as axial developments around which a fundamental transformation of modern society takes place. Equally often, other scholars have questioned and debated (i) to what extent informational processes and information-based sectors are really transforming modernity's constitution and outlook; (ii) the sometimes overly optimistic or structural pessimistic evaluations of such developments for society; and (iii) the technology determinism that follows from some of these analyses. From our analysis of the Information Society and Information Age literature over the past forty years, a number of conclusions seem relevant for this volume. First, although in the 1970s and 1980s there was considerable debate as to what extent information technologies and information processes were really changing the constitution of modernity, at the turn of the millennium – with new information and communication technologies spreading quickly and widely, and accelerated processes of globalisation – this debate seems to be less intense. Increasing consent seems to emerge on the significant influence of informational developments in restructuring the modern order. Second, more recently, within the Information Age literature, a set of new issues has emerged, among which structural uncertainties (instead of rather unproblematic knowledge generation), the role of states in global governance (compared to a strong nation-state perspective) and global networks and flows (instead of just information technologies). Third, throughout these forty years, the environment has remained largely absent in the theoretical studies that try to understand how and to what extent informational developments transform modernity. Limited attention has been given to the consequences of these perceived informational developments for both the environment directly and environmental governance and reform.

The next chapter starts at this latter end by analysing to what extent information and informational developments have been included in the ideas, reflections and debates on environmental governance and reform over the past forty years.

3 | Social theories of environmental reform

1. From environmental crises to environmental reform

The Information Society literature and the studies on the Information Age paid only marginal attention to environmental questions. During the 1970s and 1980s, this 'environmental omission' can be seen as representative for the wider social sciences, rather than an anomaly of this particular literature.[1] But 'the environmental question' was not completely absent from the social sciences. During the late 1960s, and especially the 1970s – that is, parallel in time to the emerging Information Society debate – several social sciences witnessed the emergence of relatively small environmental subdisciplines: within sociology, political sciences, economics, and later also within anthropology and law. Strongly triggered by social developments in Western industrialised societies, social scientists started to reflect on a new category of phenomena: the changing relations between nature and society and the reflection of modern society on these changing relations.

In retrospect, the framing of environmental questions within sociology and political sciences during the 1970s and 1980s was of a particular nature. The emphasis was primarily on the fundamental causes of environmental crises in Western industrialised society and the failure of modern institutions to adequately deal with these environmental crises. Environmental protests and movements, state failures, the capitalist roots of the environmental crisis, and environmental attitudes and (mis)behaviour were the typical subjects of environmental sociology and political sciences studies in the 1970s. Many of these studies were strongly related to neo-Marxist interpretation schemes (cf. Enzensberger, 1973; Schnaiberg, 1980; Pepper 1984), and even

---

[1] See, for sociology, Catton and Dunlap (1978a; 1978b), Buttel (1978) and Spaargaren (1987); see, for anthropology, Milton (1994).

today neo-Marxism is a powerful and far from marginal explanatory theory in environmental social science research.[2]

It is only by the late 1980s, and especially in the 1990s, that attention in environmental sociology and political sciences started to change somewhat toward what the sociologist Fred Buttel (2003) has labelled the sociology of environmental reform. Strongly driven by empirical and ideological developments in the European environmental movement, and by the practices and institutional developments in some 'environmental' states, European sociologists began reorienting their focus towards environmental reforms (only later and sometimes less strongly to be followed by U.S. and other non-European environmental social scientists). In this chapter, I will review these social science contributions to understanding environmental reform, by focusing on three generations of social theories, and conclude on their 'informational dimension'.[3] Although these three generations have an historical dimension in that each has been developed in a specific period (and geographical space), they are not mutually excluding or full alternatives. First-generation theories on policy and protest are still applied and relevant today, be it in a somewhat different mode as initially developed in the 1970s. In addition, insights from the first-generation theories have often been included in reform theories of later generations.

As such, this chapter can be read as the counterpoint to Chapter 2. Chapter 2 started from the Information Society and Information Age literature and related them to perspectives on environmental reform; this chapter will commence with exploring the sociological and political science perspectives on environmental reform and will explore to what extent these have been combined with information, information technologies and information flows. As in Chapter 2, we will

---

[2] Arguably, this currently is more the case in the United States than in European countries. For a comparison between the developments of U.S. and European environmental sociology (including the position of neo-Marxism), see Mol (2006b).

[3] It goes without saying that such a focus and emphasis on environmental reform studies/approaches does not disregard other environmental social sciences traditions (e.g., attitude-behaviour paradigms; political economy views; social constructivist perspectives; cultural theories) as being irrelevant. Several of these other perspectives are drawn into our analysis later in this chapter, or will appear later in this volume. It is basically through the nature of this volume that environmental reform studies are taken as the main entrance into the literature.

argue and illustrate in this chapter that environmental sociology – and the other environmental social science subdisciplines – have only marginally and rather recently included information, knowledge and informational developments as central categories in approaching environmental reform.

2. First-generation theories: policies and protests

Although emerging as a more central theme in environmental sociology and political sciences only in the late 1980s, the subject of environmental reform also has been around since the early days of the environmental social sciences. In its birth days in the 1970s (cf. Mol, 2006b; Buttel, 2002; Dunlap 2002), American and European environmental sociology and political sciences dealt with environmental reforms predominantly via two lines: analysing national environmental policies and environmental state formation and studying environmental NGOs and protests.

As environmental problems and crises were mainly conceptualised as (capitalist) market failures in the provision of collective goods, the emerging environmental state institutions were widely conceived as among the most important developments to deal with these failures. The establishment of national and local environmental ministries and authorities, new national frameworks of legal measures and regulations, new assessment procedures for major economic projects and other state-related institutional innovations drove sociological and political sciences interests, analyses and investigations towards understanding environmental reform processes. To a significant extent, these analyses were sceptical of the nation-state's ability to 'tame the treadmill' (Schnaiberg, 1980) of ongoing capitalist accumulation processes and related environmental deterioration. Building strongly on neo-Marxist analytical schemes, the state was often perceived to be structurally unable to regulate, control and compensate the inherent environmental side effects of an ongoing capitalist accumulation process. The environmental crisis was seen as being closely and fundamentally related to the structure of the capitalist organisation of the economy, and the 'capitalist state' (Jessop, 1990) was considered to be unable to change the structure of the capitalist economy. Jänicke's (1986) study on state failure accumulated many of the insights and themes of this line of investigation. Notwithstanding this dominant position during the birth period of environmental sociology and political sciences, some did see and analyse the environmental state as of critical importance for environmental reform, for instance, from a tragedy of the commons/free-rider perspective, a more applied policy science analysis, or a Weberian rationalisation view. Much research was normative and design-oriented, focusing on the contribution to and development of new state-oriented institutional layouts for environmental policy and reform. Environmental Impact Assessment schemes, environmental integration models, policy instruments, control and enforcement arrangements and the like were typical subjects for agenda-setting and implementation research.

Environmental nongovernmental organisations and civil society protests formed a second object of early environmental social science research on environmental reform. Investigations into local community protests on environmental pollution and studies on local and national environmental nongovernmental organisations constituted the core of this second branch of environmental reform analyses in the 1970s and early 1980s. The resource mobilisation studies in the United States (cf. Zald and McCarthy, 1979; McCarthy and Zald, 1977a; 1977b; 1980) and the new social movement approach in Europe (cf. Offe, 1985; Klandermans, 1986) were two dominant perspectives among a wide range of studies that tried to understand the importance of civil society in bringing about social transformations in the core institutions of modern society. In addition to a clear emphasis on the protests against what were seen as the fundamental roots of the environmental crises (Pepper, 1984), many studies also focused on the contribution of the emerging environmental movement to the actual and necessary reforms of the modern institutional order, be it via escapism in small communities detached from the dominant economic (and often also political) institutions (cf. the small is beautiful postindustrial utopians; Frankel, 1987); via public campaigning against polluters; via lobbying and influencing political processes; or via awareness raising and attitudinal changes of citizens and consumers. Among environmental sociologists there was often a significant degree of sympathy with, and even involvement in, these new social movements. Many of the more radical and structuralist analyses of the 'roots of the environmental crises' saw – and still see – the environmental movement as the last resort for bringing about change and reform.

Arguably, one could even add a third category to environmental reform studies in the 1970s, be it that this category was stronger psychology – instead of sociology or political science – based: research on environmental values, attitudes and behaviour. Strongly rooted in psychological models and theories a new line of investigation developed in the 1970s, relating changes in environmental values and attitudes of individuals to behavioural changes. Ajzen and Fishbein's (1975; 1977) model of reasoned action formed the basis for much fundamental and applied research, trying to relate polling and surveys on environmental values with concrete environmentally (un)sound behavioural actions and changes in social practices. In sociology, Dunlap's (1980) dichotomy of Human Exemptionist Paradigm (HEP) and the New Ecological Paradigm (NEP) formed a strong model for survey research, although it was initially developed to criticise the mother discipline for failing to take environmental dimensions into account in explaining social behaviour.

Reviewing in retrospect these contributions to social science research on environmental reform, one can draw several conclusions. First, with Fred Buttel (2003) one can conclude that in the 1970s and 1980s the majority of the environmental social science studies were not focused on explaining environmental reform, but, rather, on understanding the continuity of environmental degradation. Second, among the relatively few environmental reform studies conventional political and civil society institutions received most attention, whereas economic institutions and organisations, or mixes (hybridisations) of institutions/organisations, were almost absent. This was, of course, related to the actual state of environmental transformations in OECD countries during the 1970s and 1980s. Third, although during that period neo-Marxist perspectives dominated the sociology/political sciences of environmental devastations, no clear single dominant theoretical perspective emerged among the variety of environmental reform studies. Fourth, although these traditions in studying environmental protest, politics and attitudes originate in the 1970s, they still have a strong position in contemporary social sciences research on the environment as the programs of, for instance, the annual, two-yearly or four-yearly conferences of, respectively, the American (ASA), the European (ESA) and the International Sociological Association (ISA) illustrate.

3. Second-generation theories: ecological modernisation

From the mid-1980s, but especially since the early 1990s, an explosion of empirical studies have emerged on environmental improvements, ecological restructuring or environmental reform. These studies have focused on distinct levels of analysis: individual producers, households or social practices; industrial sectors, zones, chains or networks; nation-states or countries (e.g., Jänicke et al., 1992; Jänicke et al., 1997; Jänicke and Weidner, 1997). These studies all tried to assess whether a reduction in the use of natural resources and/or the discharge of emissions can be identified, either in absolute or in relative terms, compared to economic indicators such as GNP. This development is manifest in studies on cleaner production, industrial metabolism or industrial ecology (cf. Ayres and Simonis, 1995; Fisher-Kowalski, 1996; see also the *Journal of Cleaner Production*; the *Journal of Industrial Ecology*), investigations on dematerialisation and factor four/ten and perspectives on the greening of consumption, lifestyles and households. Although most of these empirical studies emerged in developed OECD countries, many of them have – be it often a little later – also found their way to less developed parts of the globe (especially Asia and Latin America).

Although not all of the conclusions in these studies point in the same direction, the general picture can be summarised as follows. From the mid-1980s onward, a rupture in the long established trend of parallel economic growth and increasing ecological disruption can be identified in most of the ecologically advanced nations, such as Germany, Japan, the Netherlands, the United States, Sweden and Denmark. This slow-down is often referred to as the decoupling or delinking of material flows from economic flows. In a number of cases (regarding countries and/or specific industrial sectors and/or specific social practices and/or specific environmental issues), environmental reform has even resulted in an absolute decline in the use of natural resources and/or in discharge of emissions, regardless of economic growth in financial or material terms (product output). These conclusions are sometimes also valid for rapidly industrializing and modernizing countries in, for instance, Asia (e.g., Sonnenfeld and Mol, 2006).

The social dynamics behind these changes, that is, the emergence of actual environment-induced transformations of institutions and social practices in industrialised societies, became one of the key objects of social science research in the 1990s. Although different concepts have been used by scholars in this field, I will group the studies that try to understand, interpret and conceptualise the nature, extent and social dynamics of environmental reform processes in this era under the label of ecological modernisation.[4]

Fundamentals of ecological modernisation

The basic idea of ecological modernisation is that, at the end of the second millennium, modern societies witness a centripetal movement of ecological interests, ideas and considerations in their institutional design. This development crystallises in a constant ecological restructuring of modernity. Ecological restructuring refers to the ecology-inspired and environment-induced processes of transformation and reform in the central institutions of modern society.

Within the so-called Ecological Modernisation Theory this ecological restructuring is conceptualised at an analytical level as the growing autonomy, independence or differentiation of an ecological rationality vis-à-vis other rationalities (cf. Mol, 1995; 1996a; Spaargaren, 1997). In the domain of states, policies and politics an ecological rationality emerged already in the 1970s and early 1980s, and 'materialised' or 'institutionalised' in different forms. The construction of governmental organisations and departments dealing with environmental issues dates from that era. Equally, environmental (framework) laws, environmental impact assessment systems and green political parties date back to that period. The same is true in the domain of ideology and the life world. A distinct 'green' ideology – as manifested by, for instance, environmental NGOs, environmental value systems and environmental periodicals – started to emerge in the 1970s. Only in the 1980s, however, this 'green' ideology assumed an independent status and could no longer be interpreted in terms of the old political ideologies of socialism, liberalism and conservatism, as argued by, among others, Paehlke (1989) and Giddens (1994).

---

[4] A full historical analysis and overview of developments in ecological modernisation literature until now is beyond the scope of this chapter. For such overviews, see the volumes edited by Spaargaren et al. (2000) and Mol and Sonnenfeld (2000), and special issues of the journals *Environmental Politics* (2000, no. 4), *Geoforum* (2000, no. 31) and *Journal of Environmental Policy and Planning* (2000, no. 4).

However, the crucial transformation that makes the notion of the growing autonomy of an ecological rationality especially relevant is of more recent origin. After an ecological rationality has become relatively independent from the political and socioideological rationalities (in the 1970s and 1980s), this process of growing independence began to extend to the economic domain in the 1990s. And because, according to most scholars, this growing independence of the ecological rationality from its economic counterpart is crucial to 'the ecological question', this last step is the decisive one. It means that economic processes of production and consumption are increasingly analysed and judged, as well as designed and organised from both an economic *and* an ecological point of view. Some profound institutional changes in the economic domain of production and consumption have become discernable in the 1990s. Among these changes are the widespread emergence of environmental management systems in companies; the introduction of economic valuation of environmental goods via the introduction of ecotaxes, among other things; the emergence of environment-inspired liability and insurance arrangements; the increasing importance attached to environmental goals such as natural resource saving and recycling among public and private utility enterprises; and the articulation of environmental considerations in economic supply and demand, for instance by ecolabels. Within ecological modernisation ideas, these transformations are analysed as *institutional* changes, indicating their semipermanent character. Although the process of ecology-induced transformation should not be interpreted as linear, evolutionary and irreversible, as was common in the modernisation theories in the 1950s and 1960s, these changes have some permanency and would be difficult to reverse.

Some environmental sociologists and commentators in the environmental reform tradition go even one step farther. They suggest that environmental considerations and interests not only activate institutional transformations in contemporary industrial societies, but even evolve into a new Grand Narrative (cf. de Ruiter, 1988; various deep ecologist scholars). The traditional Grand Emancipatory Narratives of modernity (e.g., the emancipation of labour, the dissolution of poverty) place us in history as human beings who have a definite past and a more or less predictable future. Now that these traditional narratives have ceased to perform as overarching 'story-lines', some believe the ecology (or, alternatively, sustainability) will emerge as the new sensitising concept through which modern society orientates itself in its future development. Environment/sustainability – or, rather, environmental/sustainability considerations and interests – is then the leading notion, the structuring principle, the *leitmotiv* for a new round of institutional transformations in what can be labelled (in a variation on Hobsbawm; cf. Chapter 2) the 'Age of Environment'. That still needs to be proven, but recent (2007) societal developments with respect to climate change are for some further proof in this direction.

Ecological modernisation as environmental reform

Most ecological modernisation studies focus on actual environmental reforms in specific social practices and institutions. As I have indicated elsewhere (e.g. Mol, 1995; 2001), an ecological modernisation perspective on environmental reform can be categorised in five themes.

First, there are studies on three new interpretations of the role of science and technology in environmental reform. Science and technology are no longer only analysed and judged for their contribution to environmental problems (so dominant in the 1970s and early 1980s), but also they are valued for their actual and potential role in bringing about environmental reforms and preventing environmental crises. Second, environmental reforms via traditional curative and repair technologies are replaced by more preventive sociotechnological approaches that incorporate environmental considerations from the design stage of technological and organisational innovations. Finally, the growing uncertainties with regard to scientific and expert knowledge and complex technological systems in bringing about environmental reforms do not lead to a denigration of science and technology in environmental reform, but, rather, in new environmental and institutional arrangements.

A second theme covers studies focused on the increasing importance and involvement of economic and market dynamics, institutions, and agents in environmental reforms. Producers, customers, consumers, credit institutions, insurance companies, utility sectors and business associations, to name but a few, increasingly turn into social carriers of ecological restructuring, innovation and reform (in addition to, and not so much instead of, state agencies and new social movements; cf. Mol and Spaargaren, 2000; Mol 2000). This goes together with a focus on changing state-market relations in environmental governance, and on a growing involvement of economic and market institutions in articulating environmental considerations via monetary values and prices, demand, products and services and the like.

A third theme in ecological modernisation relates to the changing role, position and performance of the 'environmental' state (often referred to as *political modernisation* in Europe [cf. Jänicke, 1993; Tatenhove et al., 2000], or regulatory reinvention in the United States [cf. Eisner, 2004]). This theme evolved in the mid-1990s in environmental governance studies. The traditional central role of the nation-state in environmental reform is shifting, leading to new governance arrangements and new political spaces. First, there is a trend towards more decentralised, flexible and consensual styles of national governance, at the expense of top-down hierarchical command-and-control regulation (a trend often referred to as political modernisation; cf. Jänicke, 1993). Second, there is a larger involvement of nonstate actors and 'nonpolitical' arrangements in environmental governance, taking over conventional tasks of the nation-state and conventional politics (e.g., privatisation, public-private partnerships (Glasbergen et al., 2007), conflict resolution by business-environmental NGO coalitions without state interference and the emergence of subpolitics[5]). Finally, supranational and global environmental institutions and governance arrangements to some extent undermine the conventional role of the sovereign nation-state or national arrangements in environmental policy and politics. As I will outline later in this chapter, this a more than just a matter of scale; it is, rather, a fundamental change in environmental reform dynamics, in need for a different environmental sociology and political sciences.

Fourth, the modification of the position, role and ideology of social movements (vis-à-vis the 1970s and 1980s) in the process of ecological transformation emerges as a theme in ecological modernisation. Instead of positioning themselves on the periphery or even outside the central decision-making institutions on the basis of demodernisation ideologies and limited economic and political power, environmental

---

[5] As Beck explains, "sub-politics is distinguished from 'politics,' first in that agents outside the political or corporatist system are allowed to appear on the stage of social design [. . .], and second, in that not only social and collective agents but individuals as well compete with the latter and each other for the emerging shaping power of the political" (Beck, 1994: 22).

movements seem increasingly involved in decision-making processes within the political and, to a lesser extent, economic arenas. Legitimacy, accountability, transparency and participation are the new principles and values that provide social movements and civil society the resources for a more powerful position in environmental reform processes. Within the environmental movement, this transformation goes together with a bipolar or dualistic strategy of cooperation and conflict, and internal debates on the tensions that are a by-product of this duality (Mol, 2000).

And, finally, ecological modernisation studies concentrate on changing discursive practices and the emergence of new ideologies in political and societal arenas. Neither the fundamental counterpositioning of economic and environmental interests nor a total disregard for the importance of environmental considerations are accepted any longer as legitimate positions. Intergenerational solidarity in the interest of preserving the sustenance base seems to have emerged as the undisputed core and widely shared principle, although differences remain on interpretations and translations into practices and strategies.

Hence, all in all, this gives a much wider agenda of environmental reform studies compared to the 1970s and early 1980s, partly reflecting the changing practices of environmental reform in and between OECD countries.

Ecological modernisation and its critics

From various (theoretical) perspectives and from the first publications onwards, the growing popularity of ecological modernisation studies and ideas has met opposition and criticism.[6] Coming from subdisciplines that had been preoccupied with explaining the continuity of environmental crises and deterioration, such a move to environmental

---

[6] Cf. Mol and Spaargaren (2000). This growing importance of ecological modernisation perspectives is even acknowledged by its critics, who often do not challenge the analytical and descriptive qualities of this theory for West European societies but, rather, its normative undertones. Although contemporary environmental policies and reforms may indeed be 'based' on or reflect ideas of ecological modernisation, they should be criticised for that, as such attempts to solve the environmental crisis suffer from various problems, according to these critics.

reform perspectives cannot but meet (fierce) debate. The debates and criticism on ecological modernisation have been summarised and reviewed in a number of publications.[7] Here I want to categorise these various critiques and debates in three ways.

First, several objections have been raised during the short history of ecological modernisation, which have been incorporated in more recent versions of the theory. Although these objections against ecological modernisation made sense in referring to the first period of ecological modernisation studies (cf. Sonnenfeld and Mol, 2002), during the third generation of ecological modernisation approaches they are no longer adequate. This is valid, for instance, regarding criticism on technological determinism in ecological modernisation, on the productivist orientation and the neglect on the consumer, on the lack of power in ecological modernisation studies and on its Eurocentricity. Not withstanding the increased incorporation of these critiques in the majority (but not all) of ecological modernisation studies at the turn of the millennium, several scholars continued repeating them up until recently (e.g., Carolan [2004] on the productivist orientation; Murphy and Bendell [1997][8] on technological determinism; Gibbs [2004] on missing power relations).

Second, there are a number of critiques on ecological modernisation perspectives that find their origin in radically different paradigms and approaches. Neo-Marxist criticism by Schnaiberg and colleagues (2002; Pellow et al., 2000) emphasises consistently the fundamental

---

[7] For evaluations and critiques on the idea of ecological modernisation as the common denominator of environmental reform processes starting to emerge in the 1990s, see, for instance, Hannigan (1995), Christoff (1996), Blowers (1997), Dryzek (1997), Gouldson and Murphy (1997), Leroy and van Tatenhove (2000), Blühdorn (2000), Buttel (2000), Mol and Spaargaren (2000; 2002), Pellow et al. (2000), Pepper (1999), Schnaiberg et al. (2002) and Gibbs (2004).

[8] Murphy and Bendell (1997: 63), for instance, summarise ecological modernisation or ecomodernism as:

the perspective that treats the environment as another technological problem to be overcome in the pursuit of progress. To the ecomodernist, pollution is an economic opportunity for prevention and clean-up technologies and certainly not an indication of fundamental problems with the current economic system.

This summary and implicit criticism would have been adequate in the late 1980s, but this is no longer the case at the turn of the millennium when most ecological modernisation studies incorporated effectively such criticism in their writings.

continuity of a capitalist order that does not allow any environmental reform beyond window dressing.[9] Deep ecology-inspired scholars argue against the reformist agenda of ecological modernisation, as it opts for a light green reform agenda, instead of a deep green fundamental and radical change of the modern order, sometimes even towards postmodernity. Human-ecologists, inspired by neo-Malthusianism and sometimes in remarkable alliances with neo-Marxists (cf. York and Rosa, 2003), blame an ecological modernisation perspective for the neglect of quantities, not in the last place population growth and ever-growing consumption quantities. Consequently, ecological modernisation perspectives are blamed to be inadequate, overly optimistic/naive and incorrect. It is not so much that these objections are completely incorrect. From their starting points and the basic premises of these schools of thought, the points raised against ecological modernisation are internally logical, consistent and coherent. In various publications (Mol and Spaargaren 2000, 2002, 2004), however, we have argued that their focus is too narrow, limited and one-sided, by claiming that there is nothing new under the sun. Although ecological modernisation scholars would not deny that in various locations, practices and institutions environmental deterioration is still there, they object to the conclusion of these critics that no reforms can be identified in the institutions dealing with environmental challenges.

Third, and finally, there is a category of comments that is less easily either incorporated or put aside if we want to analyse and understand environmental reform in late modern society. These issues have to do with the nation-state or national society centeredness of ecological modernisation, the strong separation between the natural/physical and the social in ecological modernisation, and the continuing conceptual differentiation in state, market and civil society actors and institutions. Here it is especially the changing character of modern society – especially through processes of globalisation – that makes new, early-twenty-first-century environmental reform dynamics not always easily fit ecological modernisation conceptualisations of the 1990s. This is not too dissimilar to the fact that the environmental reform dynamics of the 1990s did not fully fit the 'policy and protest' conceptualisations of the 1970s environmental reform studies. It is especially these comments and discussions on ecological modernisation that have started

[9] See also the work of Pepper (1999), Blowers (1997) and Foster (2002).

the development of what can be called the environmental sociology of networks and flows.

4. Third-generation theories: networks and flows

Via various contributions, the second half of the 1990s witnessed the emergence of what we can now label the sociology of networks and flows.[10] The foundation of a new sociological perspective, a new social theory or even 'new rules of sociological methods' (Urry, 2003) never emerge with one publication, and also here several scholars are at its foundation. Crucial in the development of the sociology of networks and flows is the shift from states and societies to networks and flows of capital, people, money, information, images, goods/materials and the like. These networks and flows form the true architectures of a global modernity. Crucial for our analysis, this sociology of networks and flows had the Information Age literature, and more particular Castells's (1996/1997) work on that, as one of the crucial foundations.

It is beyond the scope of this volume to provide a full overview, review and assessment of the debates regarding the sociology of networks and flows. Others have done so with sufficient detail and balance.[11] Here we will especially focus on the main characteristics of this sociology of networks and flows, which are relevant to the environmental social sciences, and how this sociology (can) change(s) the agenda of environmental reform studies and perspectives. In addition to the focus on Castells's work on the Information Age in Chapter 2, we will here especially rely on John Urry's interpretation of a new sociology of networks and flows.[12]

A sociology of networks and flows

Although he judges Castells's trilogy on the rise of the network society as the best effort so far to analyse networked modernity, Urry (2000,

---

[10] Others have labelled this the sociology of mobilities, referring basically to the same innovations in sociological concepts and theories (see the overview of the sociology of mobility in Kloppenburg, 2005).

[11] See, for instance, Leydesdorff (2002), Simonsen (2004) and the various references in Mol and Spaargaren (2006).

[12] It goes without saying that there are numerous others that have contributed recently in developing such a new perspective, often each with his/her own terminology, emphasis and focus (cf. Kaufman, 2002; Kesselring, 2006; Graham and Marvin, 2001; Bauman, 2000; Rifkin, 2000).

2003) sets himself the task of elaborating and refining the conceptual apparatus as introduced by Castells. The two authors develop their analyses of time and space along very much the same track, although Urry does not make use of the dichotomy of the space of flows versus the space of place, which is so central to Castells's work. Instead, Urry offers more, and more detailed, concepts to analyse the development of social practices in terms of flows and networks. He suggests that one should approach spatial patterns in three ways or modalities, distinguishing among regions, globally integrated networks, and, finally, global fluids (cf. Chapter 2). The networks and flows in these three categories are partly social and partly material or technical in character. Urry employs the notion of 'scapes' to refer to networks in their function of sociotechnical infrastructures: "networks of machines, technologies, organizations, texts and actors that constitute various interconnected nodes along which flows can be relayed" (Urry, 2000: 35). The power of these network systems vis-à-vis human agents are related to the size of the networks, their density, their relations to other networks and so on. As 'large socio-technical-systems' these networks display dynamics that are described in terms of 'path-dependencies', 'lock-in-factors', 'sunk-costs', momentum, iteration and other concepts that figure prominently in the sociology of (large) technological systems. With that, Urry's sociology of flows leans heavily towards systems theory, with a moderate role for human agency and with nonhuman actants getting actors' qualities.[13]

The relevant innovations of the sociology of networks and flows for the social sciences of environmental reform are fourfold. First, with the introduction by Castells of the space of flows, and contrasting it with the space of place, a new kind of time-space organisation of practices is introduced that takes globalisation fully into account. Globalisation is no longer simply understood as elevating the same processes on a higher level.[14] Second, the sociology of networks and flows lifts the sharp distinction between the social and the material world, between flows of information and money and flows of material substance,

---

[13] Here, Urry comes close and refers to the French work on actor-network theory by Callon and Latour. In his more recent work, Latour (2004) seems more interested in, or at least pay lip service to, ecological questions.

[14] Ulrich Beck joins in criticizing the idea that with globalisation "es handelt sich letztens Endes um die Strategie eines 'Weiter-so' auf gehobenem Niveau" (Beck, 1997: 221) ("In the end, it is a strategy of 'more of the same thing' on an elevated level" [my translation]).

between the institutional infrastructure and the technological-material infrastructures. Within the sociology of networks and flows it is especially John Urry who – relying heavily on the actor-network theories of Latour (1987) and Callon (1980 and 1987) and on the reinterpretation of these by Mol and Law (1994) – tries to overcome (or do away with) the dichotomy of the social and the material.[15] In doing so, he goes way beyond the conventional schemes of environmental social scientists, who generally speaking remain comfortable with asserting that social systems should be seen as systems having a material base and with the recognition that material conditions do matter for social practices and institutional developments. Hybrids, actants and sociotechnical systems are the key concepts that point to and analyse the fading dichotomy between the social and the material. Third, the strong separation between the conventional categories of state, market and civil society is lifted, in favour of all kind of new emerging hybrid arrangements in between. Networks and flows, scapes and sociomaterial infrastructures, they all can no longer be understood in terms of state and markets. Hence, a new conceptualisation invades the social sciences. Fourth, ideas of governance, management and control drastically change following the sociology of flows. Especially in Urry's notion of global fluids, but also in more general ideas of nation-states losing their sovereignty and power, possibilities of governance and control are seriously questioned. Within Urry's (2003) work this is related to the emergence of complexity and the disappearance of agency, against the background of a strongly system theoretical framework.

An environmental sociology of networks and flows

In applying the sociology of networks and flows for understanding twenty-first-century environmental reform, and thus to build an environmental sociology (or social theory) of networks and flows, we cannot just rely on the work of Castells, Urry and other general – nonenvironmental – sociologists/social theorists. Their inclusion of

---

[15] Castells is more conventional in a strong separation between the material and the social: "after millennia of a prehistoric battle with Nature, first to survive, then to conquer it, our species has reached the level of knowledge and social organisation that will allow us to live in a predominantly social world" (Castells, 1996: 478).

environment in social theory is, at best, marginal (cf. Mol and Spaar-garen, 2006). And, to some extent, this new social theory of networks and flows runs counter to the same frictions environmental sociolo-gists had with earlier social theories (as was so strongly articulated in the HEP-NEP debate; Catton and Dunlap, 1978; 1978b). So, in apply-ing insights from the sociology of networks and flows to develop a sociology of environmental reform, we will combine the sociology of networks and flows with earlier contributions in the social sciences of environmental reform, most notably ecological modernisation per-spectives.

Whereas most of the flow literature in the social sciences emphasises flows of capital, money, images, information, and people (travel and migration) and analyses them from perspectives as diverse as economic development, governance and control, cultural diversity or democracy, an environmental sociology of flows focuses on an explicitly environ-mental interpretation of the flow concept. This environmental interpre-tation differs in two ways from the sociology of flows: (i) by analysing flows of information, capital, goods and persons from an ecological rationality point of view (by looking at environmental information, green products, green investment funds, sustainable management con-cepts, environmental certifications schemes, flows of environmental activists and their ideas); and (ii) by analysing environmental flows as such, that is, energy, water, waste, biodiversity, natural resources, contaminants and the like. Neither Castells nor Urry have developed so far an in-depth account of environmental change in either of the two ways. Environmental flows are mentioned in between all other kinds of 'flows' that could become or are the object of sociological analyses, and these other flows are not assessed for their role in and (potential) contribution to environmental governance, deterioration, or reform. Nowhere, however, do these authors argue that the set of material flows as commonly addressed within the environmental sciences and social sciences would deserve special social science reflection. Clouds, information, capital, people or wastes are analysed, conceptualised and understood in similar ways. The question is whether that is helpful for a full understanding of environmental reform. I think we are in need of a more specific *environmental* social theory of networks and flows, which builds on such general conceptualisations but specifies them for environmental networks and flows. Such a specific environmen-tal emphasis and substantiation also might reflect on and contribute to this emerging general sociology of networks and flows, perhaps stronger in the substantive formulation than in the formal social theory/ concepts.

In relating environment to (global) networks and flows – both in terms of environmental flows as well as in terms of conventional flows – we make conceptual space for new forms of environmental reform. Quite similar to most political economists and neo-Marxist environmental social scientists, Castells discusses inequalities and power in relation to the environment primarily in the context of a rather straightforward dichotomy: place-bounded environmental movements attempt to resist the omnipotent actors of the space of (economic) flows. The environment or nature enters into Castells's analysis mainly as negative side effects of the space of flows. In the end, Castells's view of environment and nature comes close to being but a reformulation of the conventional point of view of environmental economics ('externalities') in combination with the traditional 'protest-approach' in environmental sociology (social movements organizing resistance against modernity, as we saw in the first generation of the social sciences of environmental reform).[16] Within Castells's framework, there seems to be little room for including environment and environmental reform within the time-space dynamics of the space of flows itself as, among others, ecological modernisation scholars would have it. In their debates with political economy scholars, ecological modernisation scholars have made conceptual space for the inclusion of environmental ideas, rationalities and interests in the dominant economic practices and processes. In a more or less similar way, in the social theory of networks and flows environment and environmental protection should be articulated and conceptualised in the space of place as well as in the space of flow. Place-bound environmental resistance and protection by local NGOs and communities are sided by articulation of the environment in international trade, in Foreign Direct Investments, in global certification schemes such as ISO 14000 or Forest Stewardship Council labels, in transnational company networks, in worldwide epistemic communities (such as those around water or climate change) and so on. By interpreting environment and nature as attached to (also) the 'space of

[16] But Castells (2004) does make room for a globalised environmental movement that locates and operates networks of protest at least partly in the space of flows (e.g., the antiglobalisation or other globalisation movement), be it that their power to constitute and handle the switches, programs and codes that make a difference in the network society is marginal.

flows' rather than seeing them only or primarily as part of the 'space of place', questions and analyses of environmental governance and reform move beyond a defensive position of only 'blaming' intrusions and infringements of global networks and flows on the environment of local places. The 'space of flows' then becomes a relevant analytical category for protecting and articulating nature and environment, opening up sets of new scapes, networks, nodes and strategies for environmental reform.

Double hybridisation

Although the ecological modernisation school of thought already paved the way for less conventional interpretations of the role of political, economic and civil society actors in environmental reform, this is further radicalised in the environmental sociology of flows. Following the (global) governance literature, the state becomes increasingly replaced by a proliferation of governance arrangements that create new forms, institutions and networks for governing actors' behaviour. This transition from government to governance is based on the understanding that the political is not limited to the traditional concept of the state, in the sense of a delineated institution. Transformations of the state, new alliances between the state and other actors, new state-market configurations, and the state as only one of the many elements of global networks form all new foci of theoretical attention in the governance literature.

In understanding environmental reform from such a new perspective (or social theory) conventional conceptual and theoretical categories and boundaries are challenged. The classical distinctions among state, market and civil society actors and institutions are increasingly mixed up or blurred in dealing with environmental flows. For instance, when transnational corporations (TNCs) with a proactive environmental strategy are working in a 'low-governance-arena' (e.g., sub-Saharan Africa), they sometimes come to act as government-like agents, regulating flows from a broader perspective than just an economic perspective. We then can see market actors behaving like states. But it happens also the other way around: states buying and selling 'sinks' on international markets, competing for 'green product flows' and rationalizing their green-energy politics from a liberalisation and privatisation point of view. Finally, the sharp divisions between markets and states with their system-rationalities, on the one hand, and civil society with its broader rationality, on the other (Habermas, 1981), also seem to have lost some of their significance. Civil society actors are working increasingly (also) within – and thus become part of – the 'official' system. Here we can see environmental NGOs acting as multinational companies, trading in environmental liability or credibility (WWF), and actively creating 'sub-political arrangements' in direct negotiations between NGOs and market actors (see, for instance, Pattberg, 2005; Oosterveer, 2005). Sometimes nonstate actors fill the gaps, which are left open by official institutions that cannot keep up with the forces of globalisation (e.g., in nature conservation in developing countries; in ecolabelling of wood and fish products).[17] Consequently, such forms of hybridisation show significant continuities with (and sometimes further radicalisations of) the notions of political modernisation, regulatory reinvention and subpolitics, which prevailed in the second generation of environmental reform studies.

The environmental sociology of networks and flows emphasises and conceptualises such shifting boundaries and pays special attention to *hybrid arrangements* in the field of (global) environmental reform. Such arrangements can be interpreted in terms of specific combinations of global networks and scapes, around particular environmental flows. The relevant questions are of course where and when do we see, expect, need or want these kinds of hybrid arrangements, what are the network and scape characteristics of these arrangements (for instance, in terms of infrastructures, power, inclusion and exclusion), how these hybrid arrangements are related to globalisation, and what the consequences are of such arrangements for governing environmental flows in terms of, for instance, environmental effectiveness and democracy.

There is, however, a second manner in which hybridisation makes sense in the context of the environmental sociology of flows. With John Urry, one can argue that in sociology one of the most commonly used and cherished dichotomy, that of the social and the material, needs to be reconsidered and reformulated. In the tradition of Callon and Latour and the by now well-established Actor Network Theory (ANT) school, Urry criticises mainstream sociology – especially the structuration

---

[17] In Surinam, the NGO Conservation International runs a national reserve covering close to one-quarter of the country's surface (Mol, Mol and van Vliet, 2004). The Forest Stewardship Council and the Marine Stewardship Council labels are often quoted examples of nonstate activities in a domain conventionally run by the state (Oosterveer, 2005)

theory of Anthony Giddens – for overemphasising agency over (technological) structure in this respect. When, for example, the car-system is at (environmental) stake, the best way to make sense of the future development of these kind of systems is to conceive of them as *hybrid systems*, as systems in which material and social entities can no longer be separated in a meaningful way.

This challenging view could perhaps be neglected when working in thoroughly social fields such as labour relations, schooling or gender; but not when working in the environmental field. Since its inception, *environmental* sociology and other environmental social sciences have been struggling with society-nature/social-material interactions and the ways in which these interactions could best be conceptualised. Schnaiberg (1980) is exemplary in his arguments against the partial or total fusion of the material/natural and the social, because the social – according to Schnaiberg – is different from the natural in some crucial respects. Societies are 'dependent from' the sets of ecosystems they rely on for their proper functioning, but they do not function in the same (mechanistic) ways ecosystems do. Because the social is different from the natural, the sciences of ecology and sociology also should be kept separate, so Schnaiberg argued. Sociology – or the social sciences in general – should not become mixed up with ecology or the natural sciences. This plea for separate tasks and identities of the social and the natural sciences also can be found in Giddens's structuration theory, as it became very influential in sociology from the 1980s onward. When discussing epistemological issues, Giddens argued that 'those looking for natural science-based laws and explanations in the social sciences did not just pick the wrong platform, but were also waiting for a train that is never going to arrive' (Giddens, 1976).

With the arrival of the sociology of networks and flows, the ongoing debate in environmental sociology on the relationship between the social and the material has taken a new direction and radicalised. John Urry – also following Ulrich Beck in this respect – argues that some of the well-defined 'units of analyses' so frequently used in contemporary sociology, turn out to be valid only in relation to societies of the first or 'simple modernity' phase of development. Key concepts such as 'nation-state' or 'environment' – when used under conditions of second, reflexive or global modernity – seem to have lost most of their validity. The concept of environment or nature during second modernity can no longer be used in isolation from society, because nature or environment is 'pulled into society', as much as society is 'pulled into nature'. The concept of nature as external to society, either in the form of a sustenance-base carrying social activities (Schnaiberg) or as a sink and reservoir exploited for human progress (environmental sciences), is outdated. According to Beck, only when it is recognised that society and environment in reflexive modernity are intermingled in many diverging ways, can one make sense of the (world) risk-society as emerging right under our eyes. The formaldehyde in your kitchen, the bovine spongiform encephalopathy (BSE) risks in your food and the nuclear fallout of Chernobyl all give proof of the outdated character (or at least the limited use) of the sociological concept of 'nature' in isolation from social practices, networks, institutions and agents.

Power and inequality; inclusion and exclusion

Finally, the social theory of networks and flows changes our ideas of power and inequality. Within the social theory of networks and flows, power and inequality are no longer only related to ownership of capital, as has been the dominant view in neo-Marxist studies, nor to the state, as was the mainstream conviction in most other schools of thought. In addition to these 'conventional' categories of power and inequality, the sociology of flows defines new inequalities in terms of having relative access to, being included in or being decoupled from, the key networks and flows. Groups, persons, cities and regions with access to the core flows and located in or close to the central nodes and moorings of global networks, are the wealthy and powerful. Following Rifkin (2000), it is access to the information flows via the Internet, to the flows of monetary capital and to the skills of people moving around the world, that distinguishes the better-off people, groups, cities and regions from their marginalised equivalents. This 'access to' and 'inclusion in' concerns both direct access and inclusion as well as the ability and capability to structure the scapes and infrastructures to partially influence the mobile flows in terms of speed, direction, intensity and so on. Or, as Castells puts it: who has the power and capability to handle the switches between and the programs of the networks that matter?

In following this analytical path, an environmental sociology of flows perspective has two operationalisations of power and inequality. First, it pays attention to the conditions for access to environmental flows and to the scapes and networks that structure the current of strategic environmental flows. And it analyses in some detail the consequences for groups, actors and organisations to whom access is denied or who do not manage to establish links with the relevant global networks. Such an operationalisation would reorient conventional environmental flow studies, as currently conducted mainly from a natural science perspective (e.g., material flow analysis, industrial ecology, etc.). It also would enrich present additions-and-withdrawals studies, as power and inequality are being linked to flows in a more direct way (see the discussion in Mol and Spaargaren, 2005). Power is thought to reside in the 'additions and withdrawals' themselves, and not only in the social practices of production and consumption. Second, power and inequality in an environmental sociology of flow perspective also would relate to the flows of capital, information, images and persons that structure, condition and enable environmental reforms. The power and inequalities related to nonenvironmental and nonmaterial flows affect environmental reform trajectories. Those with access to and in (partial) control of the key economic and informational flows can be said to dominate the new informational world order, at the expense of the place-bound local actors outside the core nodes of the global networks.

5. Conclusion: information flows and environmental reform

With the emergence of a perspective of environmental networks and flows in environmental reform, information starts to move to the centre of environmental reform studies. Information flows are a crucial category in the sociology of networks and flows as originally developed by Castells, Urry and others (cf. Chapter 2) and can thus not be left marginalised in understanding processes of (failing) environmental reform in the twenty-first century.

Of course, knowledge and information has never been absent in the social sciences studies on environmental reform (as much as environment has never been completely absent from the Information Society literature; see Chapter 2). In the first-generation environmental reform studies, environmental information was interpreted as a rather unproblematic and mostly undisputed category. Scientific information on the environment and on environmental deterioration was seen as a crucial resource for the environmental movement in their struggles with states and economic elites. Rachel Carson's *Silent Spring* (1962) was, of course, the role model of scientific information for environmental campaigns. Through in-depth and detailed scientific investigations, the devastating environmental consequences of conventional development paths and production systems were put on the public and political agendas. In a similar way, systematic environmental information collection via monitoring and subsequent modelling of economic and ecological systems[18] formed a key input for environmental policy making. Scientific controversies on the environment did exist but remained largely internal to the scientific community.

Within ecological modernisation the monitoring and making visible of environmental flows and qualities was believed to be one of the crucial processes for environmental reform. Only via this visualizing of largely invisible environmental flows could environmental reform programs in terms of, for instance, taxation, command-and-control and labelling work. Initially, environmental knowledge and information was interpreted in an equally unproblematic way as during the environmental reform of protests and policies. Visualizing and articulating environmental interests via information was a key first step in environmental reform, and consequently resulted in debates and struggles on transparency and access to data, but not so much in fundamentally questioning the scientific underpinnings of environmental information. Especially through debates with social constructivism (e.g., Steven Yearley, 1991; John Hannigan, 1995; Bryan Wynne, 1996), and the in-migration of reflexive modernity debates (e.g., Mol, 1996) this 'simple' Enlightenment position started to shift within the ecological modernisation perspective. A more 'reflexive' approach was called for, especially when dealing with the role of science and technology in promoting sustainable production and consumption. Von Prittwitz (1993) and Cohen (1997) are exemplary for the scholars that have addressed the challenge to confront ecological modernisation perspectives with the debate on late or reflexive modernity. Under the condition of reflexive modernity, the ecological modernisation of production and consumption can no longer be thought of or designed in terms of undisputed facts, values, and futures. From a reflexive modernity point of view, ecological risks are no longer simply accepted on the authority of (natural) scientists, nor do scientists have a privileged position in pointing out the best or most promising route towards a sustainable future. Science and technology are indeed disenchanted and deprivileged, and this has potentially far-reaching consequences for the ways

---

[18] The MIT report to the Club of Rome (Meadows et al., 1972) was a landmark in the scientific modelling of ecological and economic systems.

in which environmental problems are defined, perceived and acted on by lay actors as well as policy makers.

The fact of science and technology being no longer undisputed and bereft of that special kind of authority bestowed on them in earlier times should not be confused with epistemological issues that explain the crucial differences that exist between the natural and the social sciences. When environmental problems are discussed, these two major – but, in principle, separate – issues are very often intertwined or dealt with simultaneously. This can be said to be the case when, for example, the 'social' (or 'constructed') character of the climate change narrative – explained in terms of different interest groups, media and environmental movements all contributing to a specific mix of policies – would be presented in a way as to serve as proof for the more encompassing (postmodern) statement that the environmental crisis is something that is 'invented' by social actors and groups whose interests are served best by making a lot of noise about particular environmental problems. Blühdorn (2000) seem to fall victim to this position in claiming that ecological rationality is nothing more than power politics and big money. What tends to be denied then is the fact that environmental problems do have a 'real' existence. They belong to the type of problems that needs to be analysed and understood not only as social constructs but also in terms of the language of the natural and biological sciences to a certain extent.

With the emergence of globalisation, information and information flows start to relate in a very different way to (global) processes and understandings of environmental reform. The relation between knowledge/information and environment becomes less and less framed in epistemological terms of social constructivism and increasingly in terms of information flows as a constructing force of social processes. With the centripetal movement of the Internet, ICT, and informational processes in sociology, political sciences and social theory, information also should be reconceptualised and given a new position in the third generation of environmental reform studies. How do information flows and informational processes reconfigure, restructure and govern social processes and dynamics of environmental reform in the twenty-first century? With these questions, we touch on the core subject of this volume: environmental reform in the Information Age.

4 | *Informational governance*

1. Introduction

As we saw in Chapter 2, claims on the formative powers of information and information technologies in social processes, practices and institutions have been emphasised several times during the past four decades. Most recently, Manuel Castells especially has argued powerfully that a new generation of sociotechnical information systems, parallelled by globalisation processes, has radically changed the outlook of modern society. Chapters 2 and 3 clarified that, although these informational developments coincided with ongoing calls for and practices of environmental transformation and reform, the place of information and information technologies in the sociological and political science literature on environmental reform has been marginal.

This chapter starts from the premise that we can no longer keep separated the discourse on Information Society and the Information Age, on the one hand, and the discourse on environmental governance and reform, on the other. If the Information Age literature is right in claiming that information and information technology are reorganising modern life, it will also affect processes, practices and institutions of environmental governance and reform. In investigating and understanding to what extent and how contemporary processes of environmental reform are affected by information and information technology this chapter combines the literature and insights on Information Society/Information Age (Chapter 2), on the one hand, with the political sciences and sociological studies and perspectives on environmental governance (Chapter 3), on the other. I will label the idea of increasing importance of information, information technologies and informational processes in environmental governance and reform: informational governance (Section 2). Informational governance refers then to institutions and practices of – in our case, environmental – governance that are to a significant extent structured and 'ruled' by information, informational processes, informational technologies and struggles around access to, control over, and production and use of (environmental) information. Information thus becomes a crucial element, resource and domain of power struggles in environmental governance, not only in developed OECD countries but also beyond these 'information societies'.

As in most of the recent literature on governance, the notions of environmental (cf. Davidson and Frickel, 2004) and informational governance have a positive, constructive undertone. These governance notions refer to the outlook of giving direction to social developments, to the ability of – at least to some extent – controlling and steering social processes, and to knowledgeable and capable actors and agencies that make a difference (be it conditioned by structures and systems). At the same time, governance notions contrast overly structuralist perspectives of system domination and determination. In that sense, informational governance seems to be at odds with some versions of the sociology of networks and flows, such as the systemic perspectives of global complexity.

Although, when everything is set and done, such a systemic perspective might indeed not be useful in understanding environmental governance in the Information Age, my interpretation and framing of informational governance does build on the (environmental) sociology of networks and flows and moves far away from any simple, voluntaristic idea of governance that could relatively easy (environmentally) restructure society. Structural properties of global modernity do condition, influence and structure governance in its various forms and outlooks. But also, in a complex global order, we will still have environmental governance practices and institutions, now – as hypothesised – strongly conditioned and structured by informational processes. To balance both sides, I focus in Section 3 on an ecological modernisation perspective of information flows, whereas in Sections 4 to 7 I counteract a too naive, optimist and positivist framing of informational governance of the environment (and will regularly use the term informational politics and governance). These later sections elaborate on power, uncertainties, and multiple knowledge claims, the undermining state authority and global inequalities, respectively, which all 'structure' the production and control of, access to and manipulation of information around environmental controversies.

2. Informational governance and the environment

As has been stated, information has always been of importance in environmental protection and reform. But since the 1990s, something special is at stake, with the role of information in society's attempts to protect the environment. To clarify our central line of argument, it is most useful to draw on Castells's (1996: 21) distinction between an *information* economy and an *informational* economy. According to Castells, *information economy* refers to the role information plays in economic processes. Information, in its broadest sense of communication of knowledge, has always been critical in all economies. The collection and generation of information and knowledge and the communication and exchange of information played an important role in organising markets and economic processes of production and consumption. *Informational economy*, in contrast, refers to a specific form of social organisation in which information generation, processing and transmission become fundamental sources of productivity and power. The idea of informational economy is not just referring to the importance of information in economic processes, but points at a fundamental transition of the economic order, resulting in a new technological paradigm and a new social organisation (often referred to as electronic-information-communication technology and the network society). The economy has become informational because the productivity and competitiveness of units or agents in this economy fundamentally depend on their capacity to generate, apply and process information. And here Castells, and with him various other scholars (see Chapter 2), point at a historical discontinuity. The emergence of a new technological paradigm, organised around flexible and powerful information and communication technologies and linked to processes of globalisation, results in a fundamentally different social and economic order. This Information Age is characterised by a global informational economy.

Although rooted in the economy, these transformations have, according to Castells, also significant wider consequences: for politics and governance, for culture and cultural institutions, for social movements and so on. Thus, informational politics and governance are in some respects fundamentally different from conventional politics and governance – as was elaborated in Chapter 2. It is from this insight that we develop the idea of a new informational mode of environmental reform.

To conceptually develop the notion of informational governance of the environment, we draw a parallel with the informational economy and informational politics. Similar to the role of *information* in almost all economic and political processes, information has always been of importance in environmental governance processes, almost irrespective of time and place. As information was there in the initial economic exchanges in market economies (and before), information played also its role during the birth of environmental governance in the 1960s and 1970s. And parallel to the recent emergence of *informational* economy and politics in the Information Age, we can hypothesise the recent emergence of an informational mode of environmental governance, or informational governance of the environment. The concept of informational governance then refers to the idea that information is fundamentally restructuring processes, institutions and practices of environmental governance, resulting in essentially different forms of environmental governance from the conventional modes. Where conventional environmental governance highly relies on authoritative resources and (nation-)state power, in informational governance information is becoming a crucial (re)source with transformative powers. Information and knowledge processes start to become constituting and transformative factors in environmental governance, instead of just an enabling condition for formulating state policies. Consequently, with that, the conventional environmental struggles, which have been oriented around state laws, policies and measures since the late 1960s, are increasingly relocated around access to, production and verification of and control over information. This shift in resources and locations of power struggles coincides with a shift in the dominant position of the state in environmental policy making; states have to give in as regulatory monopolists, in favour of a diversity of interdependent actors in multilevel networks constructed around flows of information.

The fact that we can articulate these changes in environmental governance does not mean a rather sudden switch away from conventional governance and fully towards new modes of informational governance. Not unlike Castells's notion of an informational economy, in which the 'old' economy continues to play a role, and the idea of informational politics in which conventional parties, political ideologies and political control to some extent continue, the idea of informational governance of the environment would not include the end of conventional state environmental governance in the foreseeable future. As we point at new tendencies, recently emerging trends and innovations in the making, we can expect to see mixes (hybrids) of continuities and changes. But the emergence of informational governance does mean that the way societies deal with and try to govern environmental challenges is changing, and these changes can only be understood if we include information and informational processes.

Constituting developments

To further understand the logic, reasons and backgrounds of the idea that informational governance is emerging, I elaborate on four wider social developments that enable, condition, strengthen and structure such a new mode of environmental governance: a new technological paradigm based on information and communication technologies, globalisation, a redefinition of the nation-state, and disenchantment with science.[1] These wider backgrounds are often not environment-specific and are furthering more generally an enlarged role of information in governance processes.

First, this new informational mode of environmental governance and reform is strongly dependent on and triggered, enabled, facilitated and structured by new information and communication technologies, which enhance capacities of environmental information generation, transmission, access and application. With the revolutionary developments in ICT, the capacity has increased among actors to collect, handle, store, spread and access (environmental) information over increasingly larger geographical scales in shorter amounts of time. It is thus not so much the substantial content of environmental information (e.g., better information, high quality information, more detailed information) that makes a difference and begins to give environmental information transformative capacities. The transformative capacity of information in environmental reform is, rather, caused by the enhanced possibilities and capacities of environmental information collection, processing, transmission and use; the increase in the number of people

---

[1] It is too simple to see informational governance as the effect, caused by these four developments. Although these developments enable, trigger, facilitate and structure informational governance, it is not a simple linear relationship of cause and effect. There is also a reverse influence of informational governance and politics on for instance the nation-state, and the further need for and development of information and communication technologies.

and institutions having access to and thus being able to make use of information; and the time-space compressing of information flows in a globalising world order.[2] The new information and communication technologies are crucial in enhancing the capacities and abilities of monitoring, measurement and collection of environmental information. ICT also helps in the transparency and making (publicly) available environmental information to ever-wider audiences by increasing numbers of individuals and organisations. And ICT greatly stimulates and eases the access, use and application of environmental knowledge and information in economic, political and social processes. Through all of these ICT-induced dynamics, information gains in 'power' and impact as a resource in structuring social practices of environmental protection and reform, vis-à-vis other authoritative, economic and technological resources.

With ICT, the timescape of information processing changes dramatically, generally towards time compression. Environmental monitoring, information dissemination, and information controversies seem to happen in increasingly shorter time, towards what Castells has labelled 'timeless time'. Automatic monitoring systems are more and more directly related to Web pages, webcams register and disseminate instantly production circumstances many miles away, information on environmental pollution scandals can travel in no time around the world, calling for and causing reactions from local communities and authorities next door and from global consumers and nation-states far away.

As such informational governance is tightly connected to a new technological paradigm, based on ICT and the Internet. The emergence of a fundamentally new, informational mode of environmental governance is, of course, not caused by only the ICT revolution. Such a claim would make our analysis rather technologically deterministic. The fact that new technologies make environmental information more widely, easily and more quickly accessible, available and processable does not automatically connote that information processes and institutions restructure environmental governance. Three other parallel – and

---

[2] It goes without saying that these variables have not the same loading for all countries (cf. Zook, 2001; Gunaratne, 2002), resulting in geographical variations in the importance and relevance of informational governance. With respect to the environment, Chapter 10 deals with informational governance in information-poor environments, focussing on two non-OECD countries.

partly interdependent – developments in late modernity coinduce the growing relevance of informational processes in (environmental) governance.

Second, there is an intimate relation between the ICT revolution and processes of globalisation, as has become evident from the various scholars discussed in Chapter 2. The geographical stretching and time compression and acceleration of all kinds of social and economic processes draw upon and are enabled by information technologies. This is not fundamentally different in (environmental) governance, as it is for economic processes. But such globalisation processes also increase the importance of information in coordination and decision making in private and public institutions and arrangements. Information flows crossing sectors and national boundaries 'fit' into globalisation dynamics, especially as some other resources prove to be much more place-bound. The deterritorialised flow character of information matches with the increasing importance of all kinds of flows and mobilities in structuring global modernity: capital, money, goods, persons and materials. These global flows and mobilities also touch on the environment: in an age of globalisation environmental change can no longer be defined in the nation-state container. The environmental profile of the new millennium is predominantly one of global (environmental) change.[3] The causal interaction patterns, networks and structures that are at the foundation of much environmental stress are fundamentally global, as are increasingly the manifestations of environmental stress. Thus, it becomes increasingly difficult to define environmental problems in terms of only place-bound localities or only at the national level (cf. Mol 2001, for a more elaborate argument about this). Consequently, environmental governance is bound up with globalisation, which changes the character of environmental governance. It is not only that the nation-state loses some of its power and monopoly position, through which nonstate-based actors, resources and rules (such as information) become more important in environmental governance (see later in this chapter). But also informational processes become globalised in various ways. Environmental information processes move beyond the level of (single) nation-states and become truly global. Increasingly, environmental information sources are global, the

---

[3] Buttel and colleagues (1992) were among the first to notice the transformation "from limits to growth to global change."

(harmonisation of) information processing is global, interpretation of environmental information is global and information circulation and consumption are global.

Third, and closely related to globalisation processes, the redefinition of states has far-reaching consequences for environmental governance, not in the last place as it jeopardises the state's long-time comfortable position as managers of public goods, as safeguarders of general interests and as protectors of the marginalised.[4] With respect to these traditional/conventional state tasks, nation-state institutions were increasingly questioned in the 1980s for their (i) deficiency in sustaining trust, (ii) loss of legitimacy and (iii) poor effectiveness. During the 1990s, states have seen their authoritative monopoly and their sovereignty being endangered, both domestically and internationally, in an increasing number of domains (including the environment). Throughout the 1990s, it became increasingly clear that state authorities were dependent on global developments as well as on nonstate actors in successfully developing and implementing policies, programs and activities. Under conditions of globalisation states lost the ability and willingness to detail the patterns, regularities and order of societies, and increasingly turned to regulating mobilities and ensuring the conditions for favourable interaction processes and flows. As Zygmunt Bauman (1987) puts it, a transformation took place from a gardener state to a gamekeeper state. These redefinitions of the state include – but are by far not limited to – the management of the environment, which has been an undisputable central task and monopoly of the state since the 1960s. Thus, top-down authoritarian state environmental policies were replaced by facilitative, participatory and consensus building processes in which states share part of their decision-making capacity with other actors. The emerging literature on governance in the 1990s and the present decade (cf. Treib et al., 2007) has focused on the diversification of actors and institutional arrangements in local, national and global arenas of governance, with the redefinition of the state in policy,

---

[4] Although some scholars interpret this as the withering away of the 'environmental state', in most industrialised and industrialising market economies, we cannot witness a sharp decrease in state institutions on the environment, in legal and regulatory activities of these state institutions, or in state capacity for the environment (except for a few incidental cases such as Russia). Or, in other words, in general OECD countries are not moving towards an environmental deinstitutionalisation with respect to the state.

management and politics as the common denominator. For one, the crucial actors involved in governance go beyond the state and include consumers, customers, NGOs, communities, media actors, producers, business associations, insurance companies and the like. Multiactors, networks, alliances, subpolitics, partnerships and a dozen of other concepts all try to catch state centrifugal tendencies.[5] Second, the dominant nation-state level regulatory and governing actions are complemented or sometimes even bypassed by supra- and subnational arrangements, in which the nation-state is still part but no longer determining the conditions, rules and pace of governance. In this subversion of nation-state governance, information flows play a crucial role. To be more precise: informational flows in governance are, on the one hand, increasingly used by the state authorities as an answer to the failing authoritative resources and the growing cross-border dimensions of governance; on the other hand, the growing informational resources empower non-state actors and make nation-state borders increasingly permeable and significantly less relevant. In that way, the emergence of an informational mode of environmental reform is closely related to ideas of multi-level and multiactor governance.[6] But these latter concepts miss part of the essence of the transformation in (environmental) governance that is at stake. New governance processes, practices and institutional designs are not only different from the older ones in that more actors and more levels are involved. It is also that the rules, resources and domains of (environmental) struggles and management change: informational flows, processes and controversies start to move to the very centre of governance. And that centripetal movement of information in environmental governance comes along with the decreasing 'control' of environmental information by the state. This dramatically democratises environmental information in various ways. All kinds of actors appear on the (virtual) stage of environmental information, whether that be regarding information generation, information transmission

[5] The growing popularity of cross-sectoral partnerships for sustainability (after the 1992 UNCED summit in Rio de Janeiro and even more after the Johannesburg summit of 2002) point to this lost state monopoly and their growing dependency on nonstate actors in environmental governance (cf. Glasbergen, Biermann, and Mol 2007).

[6] As pointed out in Chapter 1, this broad conceptualisation is strongly represented in the environmental social sciences, for instance, in ecological and political modernisation studies, in the ideas on subpolitics, and in the recent policy science debates on multilevel and multiactor governance.

or information consumption. This democratisation, transparency and easy access of a constantly growing amount of environmental information from less and less easy traceable and verifiable global sources poses new challenges to governance. Kobrin (2001), for instance, details how cyberspace and the information flows linked to that presents a unique set of new governance problems, as cyberspace is inherently transnational, and distance, place, territoriality and borders have become irrelevant. Following the idea of 'governance without government' (Rosenau and Czempiel, 1992), Kobrin (2001: 697) claims that cyberspace results in an absence of a central or 'overarching' governing authority of information.

Fourth, and finally, the emergence of informational governance is strongly related to the disenchantment with science, a tendency widely discussed in the social sciences. The fact that science has lost its automatic monopoly position as trustful and credible institution, which generates reliable and indisputable information, makes knowledge and information an object of power struggles and a resource for a wide variety of interests in (environmental) governance. Generation of information and knowledge related to governance is thus no longer restricted to or legitimately monopolised by science and scientists. The almost undisputed authority natural science and scientists once had for lay actors becomes increasingly questioned and debated, making scientific information as much vulnerable to countervailing views as, among others, information from state authorities or environmental NGOs. By that, information gains in importance in environmental struggles and it multiplied. Scientific litmus tests to distinguish right from wrong prove less and less possible, whereas the growing amount of and better access to information demands clear quality labels on environmental information. This further relates to environmental and risk uncertainties of an 'ever' globalising and increasingly complex society. With decreasing abilities of conventional institutions (of science and the state) to solve or manage these uncertainties by scientific litmus tests and quality labels (see also Chapter 6), other mechanisms appear on stage. Accountability, transparency and openness, participation, procedural requirements and social and political processes of coalition building are some of the information-based mechanisms that replace science in legitimising decisions and building trust, as we will see throughout this volume. Informational and knowledge processes play thus a key role in dealing with uncertainties, in (re)building trust and in legitimising activities and decisions in a post-positivist science period. Thus, the increase – and increasing role – of information flows and information processes are not only a consequence of growing uncertainties caused by the disenchantment with science, but also a cause.

In conclusion, the growing role of information and informational processes in environmental governance is strongly embedded in a number of wider social developments, which make informational governance tendencies no 'accidental' one-night stand – they are there to stay.

Informational governance and informational regulation

Within the literature on environmental governance, various scholars have emphasised the growing importance of information in environmental regulation and reform, especially vis-à-vis conventional regulatory instruments and strategies of governance. The growing application of voluntary agreements, of education programs, of product information and labels, and of disclosure of environmental information have received considerable attention through case studies and more general analyses. And we will use these contributions in substantiating the claim of an emerging informational governance in Chapters 5 to 10. Arguably, most close to the notion of informational governance comes one particular school of thought, which has brought several of these cases and experiences together under one common denominator: informational regulation. Particularly in the United States (e.g., Konar and Cohen, 1997; Kleindorfer and Orts, 1999; Sunstein, 1999; Case, 2001; Tietenberg and Wheeler, 1998; Karkkainen, 2001) – but also beyond that (e.g., Dasgupta and Wheeler, 1996) – informational regulation has been coined to refer to the growing importance of freely available information in environmental regulation. Following Tietenberg (1998) and Kleindorfer and Orts (1999), Case (2001: 10775) has defined informational regulation as "rules requiring mandatory disclosure of information on environmental operations or performance of regulated entities to third parties. Such regulation seeks to enlist the aid of such nongovernmental forces as economic markets and public opinion either in complement to, or as a substitute for, traditional regulatory strategies of government standard setting and enforcement."

Does informational regulation refer to the same developments as informational governance? Although there are significant relations, similarities and overlap, and constructive cross-fertilisations between the informational regulation literature and what I label informational governance, the former concept distinguishes itself from the latter in: (i) a stronger emphasis on the legal foundation of information disclosure,[7] and less so on more 'voluntary' and/or nonlegal disclosures; (ii) a more narrow interpretation of information disclosure in terms of policy instruments or tools, as alternatives for strict regulatory or strict economic policy instruments; (iii) a framing of information use strongly related to the environmental state, or the relation between state and civil society (that is, neglecting informational developments in economic networks); (iv) a rather nation-state-centric perspective, instead of bringing informational regulation in relation to processes and dynamics of globalisation; (v) a less strong emphasis of informational processes in relation to the new dynamics enabled by ICT; (vi) a rather unproblematic (simple-modernity) analysis of environmental knowledge and information, instead of interpreting these as part of the disenchantment with science; and (vii) a legal[8] and economics approach to understanding informational regulation (cf. Case, 2001), rather than a political science or sociological perspective. With the concept of informational governance of the environment we aim to develop a sociological/political science perspective on the key role of information and informational processes in environmental reform, closely linked to wider developments in the transformation of the late modern order (as referred to earlier).

3. What about ecological modernisation?

Until now, I have aimed to understand the emergence of informational governance from more general developments of modernity and modern societies, without paying too much specific attention to environmental considerations and backgrounds. But the growing importance of knowledge and information in practices and processes of environmental reform also can be understood and interpreted from a more specific

---

[7] Often strongly related to the U.S. Toxics Release Inventory, following the 1984 Union Carbide accident at Bhopal in India (cf. Case, 2001; Sand, 2002), and other national and EU Pollutant Release and Transfer Registers.

[8] The contributions to informational regulation in the so-called reflexive environmental law approach especially relates to the argument developed in this book (cf. Orts, 1995; Stewart, 2001).

environmental perspective, in line with theories of *ecological* modernisation or *environmental* sociology of networks and flows. This provides us with an additional – now *environmental* – argument and logic to understand this informational mode in the field of environmental reform.

As explained in Chapter 3, the ecological modernisation literature has interpreted the changes in processes of environmental reform in industrialised societies from the mid-1980s onwards in terms of the growing articulation (or differentiation or growing independence) of an ecological rationality (cf. Dryzek, 1987; Mol, 1995; Spaargaren, 1997). In practices of production and consumption environmental considerations, ideas and interests are more and more articulated independently from other (economic, political and social) rationalities or interest, resulting in the environmental redesign of these practices, or the institutions that govern these practices. This emerging ecological rationality can and should be distinguished from a (still) dominant economic rationality in consumption and production processes.

In modern processes of production and consumption, the dominant economic rationality and interests have always primarily been articulated via markets and prices. Economic preferences, economic interests and economic power are articulated and coordinated in terms of money and markets. Ecological modernisation scholars (cf. Andersen, 1994), but also environmental and ecological economists (cf. Ekins and Speck, 2000), have worked on questions of articulation and coordination of environmental rationalities along similar lines of money and markets, for instance, via environmental taxes and other economic instruments, via economic valuation of environmental goods and services and via green niche market development. Although effective and valuable in a number of cases, in the end this is an indirect, incomplete and also criticised way of articulating environmental rationality and interests.[9] With the growing importance of information in production and

---

[9] Different kind of methodologies (willingness-to-pay, green GDPs, etc.) for expressing environmental values and preferences in monetary terms have been heavily debated, especially in those cases in which it is pretended that all environmental interests and values could be included in such monetary terms and could thus be subjected to market coordination processes. More limited and pragmatic approaches in using monetary terms to shift markets (both demand and supply) in more environmentally sound directions have proven useful and sometimes effective.

consumption processes, and with the expansion (in quantity, time and space) of information transmission, handling and processing capacity, a more direct 'methodology' comes into reach to further articulate the environment in production and consumption. The articulation and making visible of the environment via monitoring, measuring and reporting in terms of various environmental (flow) indicators forms a much more direct and complete articulation of environmental rationality. In addition, environmental information provides environmental interest with a more independent (vis-à-vis economic interests) place in production and consumption processes than economic valuation will ever be able to do. Only when we articulate environment via information – and thus independently from economic dimensions of money and markets – it becomes evident that productivity is not exclusively related to an economic rationality. The articulation of environmental interests and rationality in production and consumption processes via environmental information (rather than economic valuation) becomes possible once information starts to play a crucial role in constituting and transforming these production and consumption processes. This is in fact exactly what happens in informational modes of environmental reform. Although economic aspects of products, production and consumption are visualised, emphasised, communicated and coordinated via prices and markets, an informational mode of environmental governance and reform points at the possibilities and practices to use environmental information to visualise, emphasise, articulate, communicate and coordinate ecological interests and rationalities in products, production and consumption. It opens up possibilities for further advancements in the process of ecological modernisation of late modern societies.

To take the argument one step further, we should relate ecological modernisation processes to processes of globalisation and time-space compression. Processes of disembedding, in which production and consumption, among others, are lifted out of their local place-bound structures and systems to be reorganised many miles away in different contexts, call for symbolic tokens to build trust over large distances. With respect to economics, prices and money fulfil the need for such symbolic tokens. If it concerns environmental protection, for instance, related to global flows of (green) products, natural resources or waste when production and consumption are separated in space and/or time, environmental information is the basis for symbolic tokens that can bridge this time/space gap. Articulating environmental rationalities via informational devices and systems is then much more 'precise', complete and adequate than via economic signals or tokens. Hence, environmental protection and reform under conditions of globalisation require environment-specific symbolic tokens, and informational devices fulfil that need; ecolabels, certification systems, standards, product information systems and webcams are a few examples. Or, to put it in Castellian language, informational flows connect the environment in the 'space of flows' with the environment in the 'space of place'. In that sense, Buttel (2006) correctly noticed continuities between the ecological modernisation literature and the new sociology of (environmental) flows. With the ability to abstract and disembed environment from its sensory experiences and local contextualities and include it in symbolic tokens based on information, it can be transferred through time and space and is thus included in the space of flows.

This potential – and also actuality as we will see in the following substantive chapters – of articulating and including environmental protection and interests in the 'space of flow' has one further analytical/theoretical consequence. The environment can no longer be interpreted as only a passive place-based recipient that is constantly brutalised by global capitalism from the space of flows. It also becomes a constructive element that constitutes and greens the 'space of flows'.

With the idea of informational governance of the environment and the inclusion of it also in the 'space of flows', we have deliberately moved away from a too structuralistic, deterministic and systemic framing of environmental brutalisation in global modernity. Rather, we emphasised the innovations, new arrangements and 'voluntarist' opportunities lying ahead of us through new modes of environmental governance in the Information Age. No matter how useful and necessary such a reframing is, we should not lose sight of some of the critical issues around informational politics that were put so strongly on the agenda by, among others, Castells. If not, we run the risk of framing informational governance too much in a managerial language, not unlike some of the writers on the Information Society did some decades ago. In the following four sections, I will emphasise four major critical points of informational politics on the environment: new power relations, uncertainties, the undermining of sovereign states and global inequalities.

4. Informational politics and power

As became clear in Chapter 2, the neo-Marxist and critical theory critics of the Information Society have made a strong case of analysing both the continuities of the capitalist character of modern society and the related inequalities and power relations that thus come along with an Information Society. These valid points have not disappeared with the emergence of the Information Age, and we have to take them seriously in analysing informational governance of the environment.

With the growing relevance of informational processes in environmental politics and governance, new questions of power and inequalities emerge. Information, information production, information transmission and information access are not equally divided over the globe, nor within countries. Where Castells sees the potentials for social movements and local communities to use informational resources and networks in their environmental and other struggles (cf. women emancipation, the Zapatistas), he is well aware of the strong links between information and the dominant players and elites in the space of flows. Multinational media conglomerates, transnational corporations (TNCs), political elites and powerful state bureaucracies have a strong position in the production, distribution and control of information. A strong recent representative of this line of argument with respect to environment is Robert Paehlke (2003). In his recent book, he analyses how the emergence of what he calls electronic capitalism is frustrating the attempts of building sustainability. Especially in his third chapter, "Electronic Capitalism as Media Monolith", he outlines how electronic information has become crucial in today's capitalism, but at the same time how this very capitalism monopolises, manipulates and (mis)uses information flows for its own – narrowly defined and usually private – purposes and interests. In dissociating himself from Bell, Paehlke (2003: 78) is most clear on his position on the idea of an Information Society or Information Age:

Electronic capitalism, as a conceptualization of contemporary society, differs from Bell's postindustrial society in at least three ways. First, both are models of a society in which knowledge has become an important organizing principle. But electronic capitalism is still first and foremost capitalist – its 'axial principle' (Bell's term) remains economic growth, not as for Bell the 'centrality of and codification of theoretical knowledge'. It is decreasingly also a society of 'state or private control of investment decisions' and increasingly a society where such decisions are wholly private concerns. Second, the most important economic sector is not finance, as neo-Marxists might argue, or research or science, as Bell might argue, but media – a particular form and structuring of information and the key means of social organization, communication, value transformation, and sociopolitical control. Third, Bell did not fully see the extent to which, and all of the ways in which, information might become a mass-marketed product in and of itself.

Basically, according to this analysis, the media is right-wing, monopolised, commercial, superficial, dominant and unsustainable. Only few exceptions exist, most notably not in North America but, rather, in Europe, according to Paehlke. Others have joined Paehlke in criticising what I have labelled informational governance with respect to environment and sustainability. For instance, Bill McKibben (1993) focuses especially on the (trivial, commercial) content of information, rather than the structure of the information producers and processors, its power relations and monopolisation.

But one has to make a further refinement and distinction. In his analysis Paehlke makes a strong division between the 'old media' (newspapers, television, radio) and the 'new media' (especially the Internet and related services). Where the old media remain pretty much in control of economic and political elites, the new media seem uncontrollable and have a much larger potential to play a key role in emancipatory, democratic and sustainability agendas, especially because the production and spreading of information via Internet is more democratised. But also here we should be aware of the digital divide, the inequalities regarding access, the control of the Internet by, for instance, undemocratic states[10] and MNCs, and the growing invasion of commercialisation on Web sites and other Internet services. Informational governance of the environment – both with respect to the 'old' and the 'new' media – will not be excluded from these inequalities and power play around information. With the growing centrality of information in

---

[10] At the November 2005 World Summit on the Information Society in Tunis, fifteen countries plead – in vain – for more state control of the Internet, by strongly pushing for a new UN organisation to manage the Internet instead of the U.S.-based nonprofit organisation ICANN (Internet Corporation for Assigned Names and Numbers). Reporters sans Frontières listed these countries as "black holes of the web" and "enemies of the Internet".

environmental governance, information increasingly becomes a battle-field in environmental struggles, with unequal positions and powers. But the (unequal) power distribution in these informational struggles might be spread differently compared to the old power inequalities in conventional environmental governance. And, as has become clear from the many ecomodernist studies, inequality does not automatically mean environmental unsustainability. Hence, in studying informational governance on the environment, it is, on the one hand, too simple to just extrapolate the power inequalities in conventional environmental politics; on the other hand, we have to be aware that informational governance does not turn into a managerial model of environmental reform, without providing conceptual space for the struggles and inequalities that come along these environmental power politics. Although there is 'something new to report' in environmental governance with the emergence of informational governance, this newness is not only or automatically a celebration in terms of better or more effective environmental governance.

5. Governance under radical uncertainty

With informational governance questions on knowledge claims and uncertainties emerge strongly in environmental politics. When information becomes an essential resource and battlefield in environmental governance and politics, the question of which and whose information is considered reliable and valid moves to the fore. Social scientists are all too aware of the multiple definitions of, and locally situated knowledge on, environmental challenges. Which, or rather whose, definitions, knowledge claims and environmental information will construct environmental agendas and govern the environment? How are informational controversies being dealt with? Although I don't yet see Urry's regressive uncertainty taking hold of and paralysing environmental governance, it is clear that informational governance is much more than just a managerial process of collecting and disseminating information, followed by rational action of the addressees of information. It is – even more than conventional environmental governance – bound up with politics and struggles on knowledge claims, problem definitions, trust and power, whereas the institutions of state and science often lack the authority to unilaterally close these controversies.

Hence, with informational governance uncertainties are becoming an inherent part of environmental protection activities. And these uncertainties are a structural part of decision making at all levels, from individual consumers and household, via firms that want to green themselves, to policy and political actors and institutions.

The background and explanations of such a more critical analysis of information in environmental reform can be traced back to the Risk Society thesis, the literature on social constructivism, and the more recent complexity variants of the sociology of networks and flows, which have been reviewed in Chapters 2 and 3.

Although acknowledging the significant contributions of the Risk Society literature and the sociology of networks and flows to our understanding of how global information (and other) flows 'structure' the complexity of late modern society in dealing with – among others – environmental risks, we should not interpret this only in terms of radical uncertainty and nongovernability. The growing feelings and experiences of uncertainties and unpredictabilities, the ongoing revisions of knowledge claims, and the continuing controversies in environmental regulation come along with new informational practices, arrangements and institutions that make such uncertainties manageable and provide closures for controversies. We can witness new actors and procedures emerging on the stage of informational governance to form new arrangements and institutions in dealing with uncertainties, different knowledge claims and informational controversies. Beck's reflexive modernity analysis, for instance, has set the scene for a whole area of studies, suggestions and – to some extent – practices for alternatives to 'normal science' approaches: citizen's science (Irwin, 1995), democratic science, lay actor involvement, dialogical science, participatory decision making and so on. In a similar way, uncertainties have transformed policy and governance processes, as transparency, accountability, participation and independent verification become key issues in struggles around informational governance, instead of the state and science monopolies that characterised environmental governance in the preinformational governance era. So, instead of mourning over the loss of states and science to decide unilaterally or focus on the stalemate positions caused by these radical uncertainties of global complexities, we should investigate the institutional innovations that enable things to move on, and then assess the values, strengths and shortcomings of these new institutional arrangements.

6. State authority and postsovereignty

This brings us to a third critical issue involved in informational governance, one that is closely connected to a changing position of the environmental state. Although, on the one hand, informational governance responds to problems of state governance in an era marked by globalisation, on the other hand, and at the same time it may contribute to a further undermining of nation-state institutions on the environment, of the environmental state. And as the nation-state is still seen by many environmental advocates and social scientists as the core institution in safeguarding the environment, any downsizing or marginalisation of the nation-state and state power is met with a great deal of scepticism.

The sections and chapters so far have made clear that conventional state regulation alone is increasingly ill-adapted to deal with contemporary environmental challenges in a globalised world order. But, by the same token, it remains difficult to understand how informational governance can really successfully replace the regulatory state. In that sense, the worries and warnings of 'statist' environmentalists, such as Eckersly (2004), Stevis and Bruyninckx (2006) and Jänicke (2006), against a stateless environmental governance should be taken seriously. And they should be taken even more seriously when realising that there are powerful actors with clear interests in informational rather than state regulatory environmental governance, especially if this informational governance is deprived of any teeth (teeth such as transparency, independent verification and auditing, and sanctions). Then informational governance seems to become the equivalent of a deregulatory program towards voluntary environmental governance, in which the powerful (information) players are given leeway in environmental pollution.

In an earlier study (cf. Mol, Lauber and Liefferink, 2000; Mol, 2003), we investigated to what extent all kind of new joint environmental policy-making initiatives (e.g., covenants, voluntary agreements, labelling, environmental management systems, company environmental reports and other instruments by which the role of regulatory states seems to change dramatically) should be interpreted as deregulatory strategies of business and certain states sectors, or, rather, as strategies of innovative political modernisation, leading to better adapted and more effective environmental reform. The conclusions from that study fit also the dilemmas around informational governance indicated

earlier. Four insights together explain the complexities of the environmental state in informational governance.

First, it is impossible to generalise the intentions, effects and assessments of all informational governance practices, institutions and
arrangements. Similar forms and formats of joint environmental policy making and also informational governance work out differently
in distinct political, cultural and economic contexts. Overall, conclusions independent from specific time-space constellations and contexts
cannot easily be drawn.

Second, in the field of environmental protection informational governance arrangements and practices are there to stay, similar to changing
roles and tasks of states. Both developments are rooted in fundamental transformations of modern societies and their environmental governance systems. The question should not be whether we favour them
or not, but how they are to be framed and operationalised to let them
fulfil conditions of sound, effective and democratic environmental governance.

Third, states will most likely (have to) perform significant tasks in
successful informational governance, for instance, with respect to codification of new developments, the sanctioning of transparency and
disclosures, the organising and facilitation of informational processes,
the verification of information or the verification of auditors and so on.
Similarly, (the system of) states will remain crucial institutions in setting basic environmental quality norms and standards. But one should
not be too frightened of shifts in tasks and obligations from states to
other actor arrangements, as long as the basic conditions for environmental protection remain in the end safeguarded, either by the state
or any other solid sanctioning institution. Such developments should
be judged against criteria of maintaining and providing environmental
sustainability, rather than following Pavlov reactions 'in defence of the
state'.

Fourth, it nevertheless remains extremely useful and insightful to
constantly examine and assess the origins and consequences of shifts
in governance, also those related to informational governance. What
do these shifts entail for the activities, tasks and responsibilities of the
state and of other actors and institutions? It would certainly not be the
first time that core and powerful actors manage to frame new developments for their own interests, rather than for more common interest
of environmental sustainability and democratic governance. But, by

the same token, the state and its role in environmental governance is also changing, partly as a result and consequence of globalisation and the ICT revolution. Betting and relying automatically and too strongly on a national environmental state that is no longer what it used to be might not be very helpful and effective. Empirical examples show that the power, place and role of environmental states differ according to time and place, if we compare, for instance, the strong Chinese environmental state with the weak Russian environmental state, or the strong Dutch environmental state in the 1980s and first half of the 1990s with the weak Dutch environmental state in the new millennium.

7. Global inequalities in informational governance

Finally, although processes of informational governance of the environment are transnational and linked to the global network society, they are landing at the moment only to a limited extent in many developing countries, as a result of, among others, the digital divide and poor information infrastructures; limitations in democracy, transparency and civil society; and low environmental priorities. At the same time, however, the new informational governance modes enter on these developing nations via globalisation dynamics related to transnational companies and economic networks, foreign direct investments (FDI), multilateral agreements and international institutions, a globalising civil society, and the spreading of informational infrastructures, practices and routines through various other communities (academic communities, the media, sports, tourism, etc.). Although being ill-prepared for informational governance and still struggling to get the most basic principles of conventional environmental protection installed, these developing countries are confronted with international requirements related to informational governance, on, for instance, product/process information and labelling, environmental monitoring, and reporting obligations related to Multilateral Environmental Agreements (MEAs), information disclosure programs on bi- and multilateral donor projects, and regulatory relief for transnational companies following corporate social responsibility programs. Although such pushes may stimulate transparency and democracy, they may not always be very environmentally effective and also may result in developing countries' resistance as new informational obligations are creating or fortifying inequalities in, for instance, trade and the making of new international arrangements (e.g., Krut and Gleckman, 1998). So a whole set of new questions emerge. What are the environmental governance consequences of the increasing global importance of environmental information collection, processing and accessibility for those localities, regions and practices that are disconnected from the informational highway? Are the growing global demands on environmental information processes putting developing countries again in a disadvantaged position in global environmental governance? And what does this mean for the action repertoires of international and bilateral donor agencies, of the transnational companies that work in these informational peripheries, and the global environmental movement organisations that are connected to the global informational flows and networks?

Consequently, against the background of emerging systems of informational governance in an unequal world, we have to assess (i) to what extent informational governance has already any relevance for environmental governance outside the OECD countries; (ii) how international and global informational governance arrangements and practices affect domestic environmental governance systems in developing countries; (iii) how the (power) positions of developing countries in international and global governance systems are changing under the emergence of informational governance (and not only vis-à-vis other states, but also vis-à-vis international institutions, transnational companies and global civil society organisations such as Conservation International and the WWF).

8. Conclusion

This chapter has discussed the idea of an emerging practice of informational governance in the field of environmental protection. Informational governance as an ideal-typical model differs from conventional environmental governance in that informational processes, resources and struggles move to the centre of environmental governance and politics, increasingly replacing authoritative resources, nation-state power and conventional bureaucratic processes. Informational governance finds its origin in various – mutually dependent – developments that have come together in embarking on a new millennium. The stronger role of information and communications technologies enable information to play a larger role in all kind of social, political and economic processes. Globalisation has compressed time and space in social processes and pressures states to look for new modes of governance. At the same time, conventional institutions of state and science have lost their undisputed monopoly position in dealing with (environmental) controversies. All these developments have been supportive or facilitative for informational governance practices, and they contribute to our understanding why information, knowledge and informational processes become increasingly important in environmental governance and politics.

But, at the same time, these developments do not determine or define informational governance, in terms of its modes, forms, effects, geographical reach, winners and losers, and successes and failures. In that sense, informational governance and politics is not a next step in an evolutionary process that is unfolding automatically. It is bound up with struggles, debates, experiments and diverging interests, which all make the concrete outlooks, faces, modes and forms of informational governance not only contingent but also time-space specific. Hence, four key dimensions (power, uncertainties, state authority and global inequalities) have been elaborated to delineate the contours of the informational governance playing field, and thus the 'degrees of freedom' in forms and modes that informational governance can take.

In exploring actual outlooks of contemporary informational politics and governance, the next six chapters will be less theoretical in nature and more substantive. They will focus on concrete modes, examples, dimensions, and debates of informational governance in practice. What do contemporary informational governance practices look like? What role do various actors (business, media, states, NGOs) play in environmental information, collection, handling and strategic use? What tendencies can we witness in informational governance in various sectors? Is there a new role to play for the media, now that information is moving to the centre of environmental politics? How do these informational developments with respect to environmental governance and politics differ among countries, particularly the developed and developing? Can we say anything on how these new modes of environmental governance relate to environmental sustainability? It is these kinds of questions that will guide Part II of this volume.

Praxis

5 | Monitoring, surveillance and empowerment

Our analysis of actual changes in environmental governance under new informational conditions starts with processes of environmental monitoring and information collection. This chapter focuses primarily on environmental monitoring, and the following chapters are dedicated to the use of such informational resources in governance (although, as we will see, a very strict separation is not really possible). After introducing conventional ideas and modes of environmental monitoring, attention will turn to informational innovations in monitoring and the new questions that come to the fore: questions on informational power through enhanced monitoring capacities, questions of surveillance that come along with informational power and the enhanced possibilities of countersurveillance by citizen-consumers.

1. Conventional environmental monitoring

Traditionally, nation-states and national environmental authorities have been given the task and responsibilities for collecting, handling and disseminating environmental information. This task has always been connected to the collective good character of the environment. Initially, states took the initiative to set up nationwide environmental monitoring programs, consisting of three main parts. First, states were central in measuring environmental quality of, among others, surface water, ground water and drinking water; of (especially outdoor) air quality; of pollution levels of soil (on agricultural land, industrial areas, waste dumps, and so on); and of the quality of nature reserves and parks, biodiversity and ecosystems. Second, national and local environmental authorities played a central role in collecting information on various emissions and factors that influenced environmental quality. Monitoring and measuring – directly or indirectly – of, among others, industrial and agricultural emissions to air, water and soil, domestic emissions of waste, waste water and noise and emissions of traffic and energy systems, were seen as a key responsibility of state authorities. Third, from the 1970s onwards – and with respect to food products even much earlier – national authorities constantly monitored product qualities and various toxic elements in them. Food products were of particular concern, because of their direct relation to human health. But also toxic substances in other products were included in systematic monitoring programs, such as heavy metals in plastics and paints. In the 1970s and 1980s, these monitoring programs by state authorities primarily were related to legal requirements and control and enforcement activities. The collected data was used internally by state agencies for assessing and evaluating environmental policies, for enforcing legal requirements, for identifying new policy priorities and for updating policy plans. Consequently, most of the raw environmental data collected were kept within the state institutions, and little was released and reported to the public. If published at all, environmental data were included into scientific monitoring reports, national environmental plans, or state-of-the-environment publications, accessible and understandable for a small inner circle group of environmental experts from state and nonstate organisations. Often such publications lacked raw data, were aggregated and could not be traced back to original sources or specific localities.

Although these measuring and monitoring activities were considered to be primarily a task of state authorities, more than incidentally the actual collection of data was granted to parastatal or independent scientific and research institutions, although the monitoring remained state-organised and state-financed. Data was collected by these (state and scientific) institutions mainly in three ways. First, a very small part was collected by actual firsthand monitoring and measuring programs and activities. In most OECD countries, (automatic) air quality monitoring stations were set up in the 1970s and 1980s, first in the most polluted areas and later countrywide. These stations (automatically) collected on a monthly, weekly, daily or even hourly basis air quality data according to a number of parameters. In addition to these routine monitoring programs incidental measurements took place on specific targets (polluters, sites, areas, product categories) regarding specific parameters. More or less similar programs and activities took place with respect to other environmental media (water, soil, biodiversity), with respect to specific emission sources (industries, water treatment plants, energy production units, etc.) and with respect to product categories (food products, durable products, toys, etc.). Second, these monitoring institutions collected information indirectly via polluters or intermediary organisations. In a number of cases and countries, industrial and agricultural polluters and utility companies were legally required to send in information on emissions and resource use to state or parastatal organisations, linked with various strategies to control and verify reliability and completeness of the data. Third, a significant part of environmental data and information consisted of indirect information calculated via models, processing schemes and extrapolations. Based on known parameters environmental data were estimated and calculated, in order to fill the gap of primary environmental data. Financial costs, legal constraints, lack of sufficient and high-quality personal and equipment, political unwillingness and the sheer size of the monitoring task were behind the rapid developments in data collections via modelling, calculations and estimations.

Ever since the start of systematic environmental measurements and monitoring (organised) by state authorities, there has been a parallel circuit of data collection and monitoring by independent scientific institutions and environmental nongovernmental organisations. From the start, the environmental monitoring activities of independent scientists and environmental NGOs differed in at least three ways from the official state-organised monitoring programs. First, the state run environmental monitoring programs were – or at least had always the intention to be – more systematic, routine and comprehensive, whereas those of environmental NGOs were rather incidental, project-based, short-run and focused on one or a few parameters. Second, the monitoring efforts of environmental NGOs and independent scientists were usually part of a countervailing strategy, either by filling information gaps that the state left behind or by presenting contradictory information to pursue state authorities and polluters into clean-up actions. Third, environmental data and information collected by environmental NGOs and independent scientists were generally more accessible and open for a wider public than state environmental information. Often, the main goals for information collection and measurement by the former was transparency, public access and dissemination, in order to generate sufficient public pressure on the powers that be. Also in these days, (environmental) information was seen as a strategic resource in environmental conflicts and obtaining and publishing environmental information developed into a key strategy of environmentalists.

2. Innovations in monitoring arrangements

Basically, this system of environmental monitoring is still intact in most OECD countries and in development in many developing countries (cf. Chapter 10). There is a constant process of further refinement and improvement of basically the same monitoring and information-generation practices and institutional designs, for instance, by including more locations, more polluters and more substances and by putting more emphasis on precautionary information generation (Cranor, 2005). But as informational processes and informational resources become increasingly crucial in processes of environmental governance throughout the 1990s, also new constellations of monitoring practices and arrangements emerge, with a new set of questions. The most significant changes in environmental monitoring practices and arrangements under conditions of informational governance can be summarised in five points.

First, the scales of environmental monitoring are changing dramatically compared to, say, two decades ago. Monitoring is increasingly globalised. Examples are the monitoring systems related to advanced model-based global information on greenhouse gasses or the global monitoring by environmental satellites.[1] But also monitoring of deforestation and of wildlife – spreading, behaviour and even reproduction – makes increasing use of global satellite and positioning systems.[2] This increases the capacity to get on-the-ground data from anywhere at any time, against increasingly lower costs. According to some, we are approaching a time when virtually all pollutants will be susceptible to tagging, tracking and measurement at relatively low costs (Allenby et al., 2001). Although the construction of these global positioning and data-collection systems is monopolised by northern actors, access and use of these systems is less so but still runs along lines of the digital divide, putting critical issues of informational democracy to the fore. At

---

[1] See, for instance, http://www.temis.nl/ for global satellite-based monitoring of real-time air pollution. The various Geographical Information Systems (GIS)-based systems of monitoring and data reporting are also widening in geographical scales. See also Richards et al. (2001) and Hansen (2000) for the increasing possibilities of environmental monitoring by combining satellites and remote sensing.

[2] Such as the recent use of satellites in monitoring the positioning, movement and reproductive behaviour of giant panda bears in China by Chinese researchers and authorities.

the same time, we witness that although significant parts of the actual monitoring itself are carried out at a national level, the standardisation, the computation of data, the availability of information, and the active spreading is increasingly transboundary organised. Global environmental data sets of UN organisations (United Nations Environmental Program [UNEP], World Bank, Food and Agricultural Organisation [FAO], United Nations Development Program [UNDP], United Nations Industrial Development Organisation [UNIDO] and others), private research institutes such as the World Resources Institute (WRI) or Worldwatch Institute, and regional institutions such as the EU and its European Environmental Agency are increasingly becoming available. This multidimensional globalisation in monitoring is parallelled by a clear tendency in monitoring of zooming in on micro-processes. With ever more advanced techniques monitoring programs increasingly are able to follow and detect micro-elements at molecular and submolecular levels. Nanotechnologies, genetic modifications and other developments come together with monitoring devices and programs at similar scales.

Second, there is a change in the timescapes of monitoring, strongly enabled by ICT developments. Time-consuming sampling techniques of polluted water and air, analysing them in laboratories and publishing these results in a limited amount of hard-copy reports for state agencies seems increasingly outdated (although, as we will see in Chapter 10, this is less so for some developing countries). Continuous automatic monitoring with instant interpretation and release via digital systems and Internet seem to become the standard in OECD countries. Automatic monitoring stations for ambient air and water quality deliver data twenty-four hours a day to Web sites on the Internet, where such data and information are directly and constantly available for ever-larger groups of concerned citizens, private and public organisations, and experts.[3] In the monitoring of emissions, natural resource consumption and depletion, and products the time lag between actual environmental disturbances and the making available of information on those disturbances to relevant actors is diminishing. In addition, the increasing and advanced use of monitoring via computer modelling

---

[3] For example, wildlife is increasingly monitored in real time by Webcams, available to the wider public. See, for instance, the monitoring of a Falcon breeding nest in spring 2006 in the Netherlands, made available via the Web site http://www.planet.nl/planet/show/id=2219209/sc=cf508f.

greatly enhances the amount of data generated, but also the instantaneous availability of data.

Third, there is a clear development from making point source data available in numerical tables towards the combination of environmental information with geographical mapping. Increasingly, monitoring information is available and presented via geographical information system maps on Internet, often in combination with data presentation formats that make environmental information accessible and understandable also for nonexperts. A wide variation of models, formats and presentation techniques appear. There is a clear tendency to link this with increased traceability of pollutions to the geographical sources or – if it concerns products – to the origin in product chains, putting more emphasis on the dynamic environmental flows than on only the static sinks.

Fourth, there is a clear tendency to include a larger number of actors in monitoring on three fronts: with respect to monitoring agencies, with respect to actors that have access to monitoring data and with respect to actors that are monitored from an environmental point of view.[4] The strong monopolisation of environmental monitoring activities, of environmental data ownership, and of data dissemination control by the state and economic powers seems to have gone. "One of the remarkable stories behind the Information Age is how much environmentally relevant data and knowledge is being generated and shared without any plan, governmental mandate, or structured set of incentives" (Esty, 2004). In that sense, one could speak of a diversification and democratisation of environmental monitoring. Environmental monitoring is no longer the privilege of national governments and large (transnational) companies. Consumers, citizens, customers, insurance companies, nature lovers, public transport users and others are regularly involved in monitoring of various activities (production, products, mobility, services, utility provisioning, investments, trade, etc.) regarding numerous environmental qualities. Although there is clearly a democratisation taking place in monitoring activities, at first sight this seems to be also the case with respect to access to monitoring data.

---

[4] Matthews (2001; as quoted in Esty, 2004) foresees a time when humans are equipped with a digital skin that records in detail the pollution impacts they are facing. There are more of these – at first sight, science fiction–like – ideas that in the end prove not to be too far from reality.

Via ICT and the Internet, access to data and environmental information is expanding. But, by the same token, we also can witness new lines of inequality in access to environmental information, along with informational governance, as will become clear when we analyse the media, the digital divide and developing countries in Chapters 9 and 10.

Fifth, the means of monitoring are also diversifying. Originally monitoring was largely related to taking physical samples and analysing these samples in laboratories in order to reveal the qualities that were not directly obvious for the senses.[5] Nowadays, monitoring is also carried out, for instance, by webcams, by satellites and by sensory experiences of lay actors. Environmental data do not only take the form of chemical and biological quantitative numbers in tabular form, but also are expressed as stories on weblogs, as pictures on Web sites, as survey data in social science studies on perceptions, as complaint systems via hot lines, or as consumer preferences on nature reserves. This is, of course, directly related to the lost monopoly of scientific data, as discussed in the previous chapters, and the subsequent revaluation of a variety of other kinds of information sources and data collection means, although Burström and Lindqvist (2002) make clear that the dominance of natural science environmental data and information still continues.

3. Who monitors who?

These developments in environmental monitoring practices in the 1990s have led to shifts in monitoring arrangements and schemes. The classical state-organised environmental monitoring of industries and other point-source polluters is diversified. In order to bring some systematics into this diversity, we can categorise contemporary monitoring schemes according to two characteristics or dimensions.[6] On the one

---

[5] Although a small part of monitoring has always taken place by using sensory experiences, for instance, when smell, natural beauty or visual pollution is at stake. Monitoring then consists of the standardisation and aggregation of individual sensory data into larger units, followed by statistical analysis along time dimensions.

[6] This categorisation, as well as the examples to illustrate the categories, center on (industrial) production and consumption practices. Although this is a significant selection covering the majority of environmentally relevant categories, some monitoring schemes do not easily fit into this, such as those with respect to nature restoration and preservation.

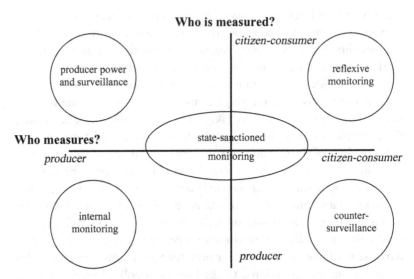

Figure 5.1. Five ideal types of monitoring schemes (following van den Burg et al., 2003).

hand, we are interested in the question of who monitors and measures: which actors are in charge of, responsible for and controlling monitoring and measuring, for example, consumption levels and practices, environmental emissions and effects, and production methods. On the other hand, we are concerned with the question of who or what is subjected to measurement and monitoring: which actors and which social practices are monitored. This results in five different ideal type monitoring schemes, which we label (1) internal monitoring, (2) producer-power and surveillance, (3) reflexive monitoring, (4) countersurveillance and (5) state-sanctioned monitoring (Figure 5.1).

Within producers/providers/companies, internal or self-monitoring of substance flows (energy, materials) and production practices is important to follow production processes, to enhance efficiency of natural resource use, and to reduce costs of waste fees, but also as a means to fulfil state requirements (emission standards), to acquire certification (e.g., reporting requirements following ISO 9000 or ISO 14001; product labelling requirements) and to enable communications with third parties on corporate responsibility and a 'license to produce'. These internal monitoring schemes are not necessarily restricted to one company, organisation or agricultural farm, but also might extend to other firms within global networks and product or value chains (related to, for instance, quality control, product liabilities and labelling schemes) or in geographical areas/regions (such as industrial parks or estates). Chapter 7 especially elaborates on monitoring and informational practices in this category, showing that a rich diversity of monitoring practices and schemes goes behind this category, triggered and pressured by various sources and factors. In the second category, producer power and surveillance, producers and other institutional actors (such as utility companies, retailers) collect environmental information and knowledge on citizen-consumers, consumption and citizen behaviour for their own use. Such monitoring schemes include, for instance, traditional utility-related metering systems, marketing investigations on (biological) products, mobility behaviour, etc. These monitoring schemes have enabled demand side management strategies in environmental governance, but the major risk of these schemes is the possible loss of privacy and increased surveillance of individuals, on which I will elaborate later in this chapter. The third category, reflexive monitoring, comprises monitoring schemes in which citizen-consumers – or their representative organisations in civil society such as environmental NGOs, consumer organisations and community organisations – monitor themselves, their households, their consumption levels and their behaviour. Monitoring then becomes part of reflexivity, in which agents reflect on their daily routines to develop alternative action patterns that address the environmental consequences of these former routines. In the fourth category, which can be labelled countersurveillance, we find monitoring schemes and arrangements in which individual citizen-consumers and (more often) consumer, environmental and other social organisations collect data on pollution levels, emissions, water quality, product quality or transport performance, to name but a few. These monitoring schemes provide citizen-consumers with the opportunity, the resources and the power to step up to providers, producers or state institutions and demand and enforce better environmental performance, increased product quality and more transparency. We can speak of countersurveillance, as citizen-consumers or civil society organisations have acquired and use the means to monitor and influence other actors (most notably producers and state authorities) more than incidentally against the wishes of these actors. In the last category, labelled state-sanctioned monitoring, monitoring is carried out, organised or sanctioned neither by consumers nor by producers, but

citizen-consumers, potentially intruding in the private lives of many citizens. At least some scholars claim that surveillance thus becomes relevant for environmental monitoring and control, emphasising drawbacks of informational governance.

Colonisation of the life-world?

In ecological modernisation theory, environmental monitoring is identified as a crucial aspect of environmental reform programmes (cf. Spaargaren, 1997; Huber, 1991). Monitoring of substance flows is the first step in environmental reform, as it reveals where interventions should be made in production and consumption systems. Increasing monitoring efforts is thus desirable, as it will enhance information and knowledge about the origins of environmental problems, which is a precondition for implementing environmental protection measures. As such, ecological modernisation scholars usually embrace the refinement and expansion of monitoring arrangements.

But monitoring social practices not only increases the amount of information and knowledge on environmentally relevant dimensions of these very practices but also generates information on the behaviour and whereabouts of the actors involved. Also with environmental monitoring the danger exists that citizens are increasingly subjected to surveillance and social control by systemic actors. In Habermasian language, this can be referred to as the colonisation of the life-world: the penetration of system elements – now via environmental monitoring by state, institutional and economic actors – into the private life of individual citizen-consumers. How serious and real is this threat under the contemporary advancements of environmental monitoring?

Monitoring can be considered as a particular form of surveillance. Dandeker's (1990: 2) definition of surveillance as "the supervisory and information gathering capacities of the organisations of modern society and especially of the modern state and business enterprises" clearly reflects this idea of control by the system. With respect to surveillance, social theories have been heavily influenced by the work of Foucault (1977), who considers surveillance to be a constitutive element of modern societies, replacing visible coercion with the sustained monitoring of conduct. Surveillance is the key to discipline, to the internalisation of rules. The Panopticon, Bentham's proposed architectural design for a prison, best illustrates Foucault's reasoning. Allowing for permanent monitoring of the inmates, the major effect of the Panopticon is "to induce in the inmates a state of conscious and permanent visibility that assures the automatic functioning of power" (Foucault, 1977: 201). According to Foucault, this kind of disciplinary power originated in prisons, asylums and the like, but subsequently migrated and penetrated into other spheres of daily life outside these restricted and guarded geographical locations (hospitals, schools, enterprises, etc.), eventually pervading the entire modern, contemporary society. With Foucault, monitoring and surveillance are evaluated in terms of power and control of institutions of the state and the economy over individual citizen-consumers.

Foucault's analysis of surveillance, power and control has been criticised for its emphasis on technocratic and monopolistic monitoring and surveillance. The construction and operation of monitoring or surveillance systems requires certain privileges, be they political or financial, and surveillance and disciplinary power therefore not only construct (unequal) relations of power, but are at the same time the product of these relations. Foucault has paid little attention to the question of how surveillance and discipline are linked to, for example, class struggle and the technical conditions of modernity. "The whole question of the relationship between interests and the disciplinary structures is pushed to the margins of Foucault's concerns" (Dandeker, 1990: 28).

A second point of critique is concerned with the rather deterministic outlook of Foucault's work. Whereas the basic principle of the Panopticon is that 'the few see the many', there has always been a strong countertendency "including the development of unique and extensive systems enabling the many to see the few" (Mathiesen, 1997: 219); both panopticism and *synopticism* are characteristics of our society. The development of mass media and the networks of civil society (whether related to environment, labour, human rights or other) are prime examples of synopticism, enabling the many to follow – to some extent at least – the actions and whereabouts of the few, of the elite. The fact that surveillance has now become so widespread, involving so many actors, also means that "surveillance has become rhizomatic, it has transformed hierarchies of observation, and allows for the scrutiny of the powerful by both institutions and the general population" (Haggerty and Ericson, 2000: 617). Although surveillance still takes place in an unequal world, no major population group or institution stands "irrefutably above or outside of the surveillant assemblage"

(Haggerty and Ericson, 2000: 618). For example, civil society–related interest groups and loosely organised transboundary networks of individuals have developed their own monitoring systems and strategies to discipline and control businesses and the cultural and political elite.

Anthony Giddens's understanding of surveillance and power seems then more suitable for analysing and understanding present developments in environmental monitoring and informational governance. Giddens sees surveillance, "the supervision of the activities of subject populations in the political sphere" (Giddens, 1990: 58), as one of the four institutional dimensions of modernity, next to industrialism, capitalism and military power. Surveillance is considered to be a structural property of both traditional and modern societies. In modern societies, surveillance reaches "an intensity quite unmatched in previous types of social order, made possible through the generation and control of information, and developments in communication and transportation, plus forms of supervisory control of 'deviance'" (Giddens, 1985: 312). Just like any other structural property or principle, surveillance does not act on individuals in a deterministic way, like the forces of nature. Monitoring and surveillance should be considered a social construction, whose functioning and social effects are dependent on the design of the monitoring and surveillance schemes and the related interests and relations of power.

This is, of course, not to say that monitoring of consumption, consumer behaviour, and citizens runs no risk of an increased colonisation of the life-world and social control. But because of the nature of the world of social actors, environmental monitoring will not automatically lead to monopolistic social control. Potential detrimental social consequences of monitoring social practices are related to the design of the surveillance and monitoring arrangements and the networks of power and interest governing such arrangements. As we have conceptualised in an earlier section in this chapter, under conditions of informational governance monitoring schemes are widening and move beyond a mere systemic monitoring, surveillance and control of the life-world. Countersurveillance, reflexive monitoring and internal monitoring are as much prevailing as monitoring arrangements as 'producer power and surveillance' and 'state-sanctioned monitoring'. In addition, we witness in late or global modernity new power relations in production and consumption chains and networks, in which post-Fordism, chain-inversion and the consumerist turn all point to more power at the consumer side of production-consumption systems (see Chapter 7). Monitoring schemes under conditions of informational governance might thus very well enhance countersurveillance and countervailing powers, rather than only colonise the life-world. We will expand on the idea of monitoring power by citizen-consumers later in this chapter.

Countersurveillance: citizen-consumer monitoring

Citizen-consumer monitoring includes self-monitoring of citizen-consumers, as well as monitoring schemes and arrangements in which citizen-consumers disclose the behaviour of producers/providers and thus develop countervailing power. Although both types are brought together as they are performed by citizen-consumers, the objectives are completely different. Reflexive self-monitoring is solely concerned with attempts to realise internal and often individual natural resource savings and waste minimisation, for either environmental or financial reasons. Such monitoring arrangements are not focused on or engaged in realisation of changes in the 'infrastructures of consumption'. Countersurveillance monitoring schemes are directed primarily on influencing companies, utility providers, production/provision practices or even state agencies, attempting to realise changes in production processes, products, infrastructures and other systemic parts.[7]

An analysis of the differences between citizen-consumer monitoring and more conventional monitoring schemes can offer more insight in the specific characteristics of citizen-consumer monitoring. The (history of) monitoring of domestic utility-related resource flows functions as a starting point to refine ideas on where characteristics of citizen-consumer monitoring differ from more conventional monitoring (cf. Marvin, Chappells and Guy, 1999; Shove, 1997; van Vliet, 2000).

---

[7] In an interesting analysis, van den Burg (2006) distinguished along similar lines horizontal and vertical empowerment of consumers with respect to monitoring. Horizontal empowerment refers then to monitoring arrangements that provide consumers further information and control over their own practices. Vertical empowerment refers to monitoring of the production-consumption chain, providing citizen-consumers with information, influence and control over providers and producers. This latter form of empowerment is closely linked to ideas of chain-inversion.

Conventional monitoring of utility-related substance flows originated in network-based systems of provision, where information on use patterns allowed producers to manage supply and network capacity. Furthermore, monitoring in energy and water networks was necessary for the introduction of individual payment systems. In some cases, citizen-consumers actually plead for the introduction of individual payment systems as they felt that they paid more than what was considered reasonable or more than their share (Vliet, 2000). Following the energy crises of the 1970s and the upsurge of environmental concern, existing (individual) water and energy meters also were considered a key technology in reducing domestic consumption of these utility-provided commodities. Monitoring technologies were thought to make the invisible visible, and through that link individual or household behaviour and consumption to prices and environmental consequences. These links were considered to be conditions and incentives for environmentally friendly behaviour by citizen-consumers, but they did not address producers and providers.

The organisation of these conventional monitoring schemes clearly reflected the underlying relationships between providers and consumers and thus to some extent determined the action that consumers were supposed to take (or not). Citizen-consumer monitoring will differ in some of its core characteristics from these conventional monitoring arrangements. Such new monitoring schemes are established with the explicit idea to target and empower citizen-consumers and to keep the explicit wishes, interests and demands of them in mind. For example, citizen-consumer energy monitoring would not only relate to what happens behind the meter (at the consumer side), but also include information of what happens at the provider/producer side: which energy source is used in producing electricity, which emissions take place, where is the power plant located, what efficiency is used? This monitoring enables and empowers citizen-consumers to press for system reorganisations. Citizen-consumer monitoring schemes could thus impact on the providers and the energy sector, for instance, by enabling (physically and financially) feedback of decentralised produced PV electricity from the roofs of individual homes into the grid.

This also relates back to the monitoring equipment and reporting practices. Regarding conventional utility meters "the precise technical configuration of the meter is strongly shaped by the often conflicting objectives of agencies involved in developing and implementing the systems" (Marvin et al., 1999: 113). In conventional or producer monitoring schemes, the collected data is given in technical units. Electricity use is expressed in kWhs, gas consumption in joules and water consumption in cubic meters. Furthermore, reporting and dissemination of monitoring results often takes place only once a year, on an aggregated basis. The bills that are supposed to give information on individual or household consumption levels are quite complex. Such monitoring and related reporting schemes are not meant to – and thus do not – provide citizen-consumers with a detailed understanding of what is going on within their households, for they do not link information on consumption of water and energy to the actual practices of consumption that are recognizable for citizen-consumers: washing, cleaning, heating, lightning, cooking and the like. In addition, these data presentations often fail to give consumers an understanding of their realised savings, possible savings per social practice and the environmental effects caused or prevented.

Thus, adequate citizen-consumer monitoring does not only differ from conventional monitoring in the subject and object of monitoring, but also in, among others, the technical devices, reporting schemes, and the type of information collected. Only then is citizen-consumer monitoring empowering.

5. Citizen-consumer empowerment

Four case studies on citizen-consumer environmental monitoring shed some light on their diversity, and how informational governance is not necessarily only linked to surveillance and system dominations, but can as much be part of countersurveillance and democracy.[8]

Energy monitoring

As introduced earlier, in most OECD countries domestic energy monitoring is done via individual metering of households by energy companies. Monthly or annual bills of utility companies feed back information on energy use from the producers/providers to the consumers. Although such systems of metering and billing can form important

---

[8] Some of these examples draw on our earlier research and publications: van den Burg, Mol and Spaargaren (2003) and van den Burg (2006).

instruments in raising environmental awareness and reducing household energy consumption, in practice several characteristics of these monitoring schemes prevent this. The technical nature of information on the bill, the lack of information on energy use related to specific domestic practices (bathing, heating, cooking, etc.), deficiencies in information on the production side of energy and the low frequency of these bills (usually once a year) give consumers little potential for countersurveillance or reflexive monitoring strategies. However, there exist other, more promising, schemes and arrangements.

Obragas, the utility company in the area of the city of Helmond (the Netherlands), carried out a telemetering experiment in 1997 and 1998. In twenty-nine households, living in newly built, energy-efficient houses, a special device was installed, which collected data from conventional electricity, water and gas meters and sent the information via a two-way TV cable to the computer network of Obragas. Consumers received a weekly personal update on their consumption of water, gas and electricity on Teletext. Actual consumption levels, aspired savings and achievements were compared and visualised through smiling, neutral or sad-looking faces. The aim of the experiment was to examine the possible contribution of regular feedback to consumers on lowering their energy and water consumption. This monitoring arrangement contributed to a water consumption reduction of 18 percent, a gas consumption decrease by 23 percent and a lower electricity consumption of 15 percent. According to an evaluation by Völlink and Meertens (1999), feedback on consumption levels was appreciated. Participants were mainly interested in comparing current consumption levels with aspired levels, rather than in the real consumption levels. Much to the disappointment of the participants, the experiment ended after three months, as it appeared to be too expensive and technically vulnerable. The permanent monitoring of household consumption by utilities was not considered a drawback by the participants (also because consumers participated voluntarily). The monitoring arrangement, however, did not allow for any influence of consumers on the infrastructures of consumption: the way that energy and water was produced and the kind of sources the utility company drew on in producing water and energy. In that sense, the project stimulated reflexive monitoring possibilities rather than countersurveillance. For utility companies, such projects are experiments in binding households to their services through information provision and extensive monitoring.

Electricity labelling and certification schemes are a form of energy monitoring in the category of countersurveillance. Energy companies are 'forced' to disclose the energy sources they rely on, how energy/ electricity is produced and how labelling takes place. In an increasing number of labelling schemes, consumers can chose different kinds of electricity, and independent parties verify whether demand in, for instance, green electricity matches production. In the EU, a system of green certificates was developed to secure that the amount of green electricity produced matches the demand and consumption levels, resulting in a market for these certificates.[9] Often (e.g., in Switzerland, some U.S. states, Austria), electricity companies also have to publish information on their electricity generation portfolio regularly in uniform standards and through multiple channels (Markard et al., 2003). As such consumers are provided information on what is happening at the producer's side. Especially through NGOs, a form of countersurveillance is built into the energy system.[10]

Global Action Plan for the Earth: ecoteams

In the early 1990s, David Gershon founded Global Action Plan for the Earth (GAP), the NGO that has developed and promotes the ecoteam program. At the core of the ecoteam program is the belief that consumers are willing to change behaviour but have insufficient knowledge on the environmental impacts of their own and of alternative consumption practices. By self-monitoring of consumption (for example, water and electricity use, waste production, products, mobility), comparing

---

[9] This leads to a further deterritorialisation of environmental flows. In the trading of these certificates, green electricity is completely delinked from its localised consumption, embedding it completely in the space of flow. Green electricity flows are no longer relevant for producers and consumers, but only the flow of certificates, or virtual green electricity. Although in the end green electricity certificates have to relate to grounded electricity production practices, and verification and audit schemes have to guarantee that, the localities of green production and green consumption can no longer be related to each other.

[10] NGOs are particularly active in reviewing electricity producers and distributors with respect to their environmental claims. The World Wide Fund for Nature (WWF), for instance, performs independent audits of green electricity schemes and Energy Watch in the United Kingdom has a similar function. In addition, various independent Web sites have made the electricity market more transparent and helped as such citizen-consumers to influence providers.

consumption patterns with a peer group, and the exchange of ideas on alternatives and innovations, consumers become more aware of their own environmental impacts and the available options to lower these. Although participants to the ecoteam approach are provided with a guide that discusses several environmental subjects and contains alternatives and environmental innovations, the key element of the ecoteam program forms the group meetings that provide participants with information, support, social pressure and incentives to change their lifestyle.

The first step of the program is self-monitoring of consumption levels. This implies, for instance, regular reading of water and electricity meters, weighing garbage, maintaining a logbook on transportation-kilometres and even registering the origin of food products consumed. Although existing meters provide knowledge in technical units and do not match the 'logics' of social practices and consumers, ecoteams prove that kWhs, cubic metres, kilometres and kilograms can become meaningful consumer experiences. Group meetings enable comparison with people of flesh and blood (as opposed to abstract nationwide averages) and frequent self-monitoring apparently teach participants to understand and interpret social practices in a technicians' discourse.

In the second step, participants are encouraged to share ideas and experiences and – together with the GAP workbook – this enables the development of tailor-made environmentally friendly alternatives. Achieved savings of group members are collected and information about the total savings of all ecoteams is regularly published in a newsletter. Evaluations from the Netherlands show that, shortly after participation, on average ecoteam participants reduce waste production with an average of 17.6 percent, gas consumption with 23.1 percent, electricity consumption with 6.8 percent, and water consumption with 4.9 percent. Six to nine months after the program had finished, consumption levels had, surprisingly, decreased even more (Staats and Harland 1995; Harland 2001).

The ecoteam program was introduced in the United States and spread to almost all countries of Western Europe, including, for example, Great Britain, Spain, the Netherlands, Sweden and Switzerland; but also to other countries, including Slovenia, Russia, Japan, and South Korea. GAP's aim is to reach about 15 to 20 percent of the population, which is believed to create enough critical mass to spread environmental friendly innovations through society. As the ecoteam program proved to be a successful method for changing individual lifestyles, several local and national governments have invested in the promotion of the program, among others by paying for regional centres.

Although the main goal of the ecoteam program is to change individual lifestyles, participants are asked to step up to utilities, companies and governments with questions, demands and comments. This confronts producers and institutional actors with consumer concerns. The information, knowledge and experiences of participants strengthen their position vis-à-vis the 'systems of provision' and may push for new consumer-provider relations. In the United States, GAP has developed the 'Livable Neighborhood Program', in which participants not only discuss individual consumption patterns but also develop actions to improve their neighbourhood, hopefully resulting in "an effective neighbourhood mobilization and action tool to assist local government in delivering services and improving the overall liveability of the community" (see GAP's Web site, http://www.globalactionplan.org).

Ecoteams can be considered one of the best developed citizen-consumer monitoring schemes. Yet the current design of the program has its limitations. First of all, one can doubt whether or not the current methods – using group meetings and social control – are suitable for targeting large parts of the population, although the success of the program – in terms of reducing the environmental impacts – seems to be dependent on exactly these methods. Second, the environmental impact of the program could be increased if the links between individual monitoring and the 'infrastructures of consumption' are tightened. In other words, if consumers gain more insight in – and develop power towards – the structures surrounding consumption, and come up with collective activities and strategies to put these structural elements under pressure, the ecoteam approach also might reach nonparticipants. But then we are reaching the boundaries of the ecoteam approach and enter in to more regular environmental and consumer NGOs (see Chapter 8).

Right-to-know: http://www.scorecard.org and others

Especially since the 1998 UN/ECE Århus Convention, the Rio Principle 10 and the Access Initiative OECD countries have sped up regulation on Pollutant Release and Transfer Registers (see Chapter 6). Strongly supported by ICT developments, this has resulted – first in the United States, but later also in other countries such as Australia, Japan, the United Kingdom, the Netherlands and other members of the EU – in direct worldwide public access to environmental emission and performance data of thousands of polluting companies. Here we will especially focus on how the U.S. environmental movement has developed the Scorecard Web site, using publicly available information (of the EPA's Toxic Release Inventory (TRI), as well as a number of other available datasets) for countersurveillance by citizen-consumers (Roe, 2002; Karkkainen, 2001; Natan and Miller, 1998; Cohen, 2000; Sand, 2002; van den Burg, 2004).

In 1998, the U.S.–based nongovernmental organisation Environmental Defense Fund developed the Web site Scorecard to provide citizens with freely accessible and understandable information on local pollution levels, polluters and possibilities to take action.[11] Entering a postal code or selecting a locality on the map provides one with an overview of the different pollution levels in that locality, the companies responsible for these pollutants and their relative ranking.

In addition to merely providing information, Scorecard also enables citizens to take action: sending a fax or an e-mail to local companies, local authorities or a congressman; joining a local environmental organisation; or starting a lawsuit. After some years, a section on environmental justice has been added to the site, with differentials in environmental burden by place, minorities, race, class and so on. Throughout the United States, Scorecard is based on existing data collected by scientific or governmental agencies. Because original data often is hard to access and difficult to interpret, Scorecard enables individuals to easily access and collect information on local environmental burdens and interpret them. This is also one of the main difficulties for Environmental Defense Fund: to collect data and translate this into a format that refers to the everyday experience of lay actors. "The best numbers available today are very far from being perfect measures of the environmental burdens that different people experiences – and of course, numbers can't tell the whole environmental justice story. But systematic data on the 'where' and 'how much' of unequal environmental conditions,

[11] In his interviews with Bill Pease (the lead developer of Scorecard at Environmental Defense Fund) and with Michael Stanley-Jones (one of the users of the Scorecard Web site when he was located at the Silicon Valley Toxics Coalition), Schienke (2001a and 2001b) provides interesting backgrounds on the development and use of this Internet- and GIS-based system.

even if imperfect, will help focus attention and set priorities in this critical area of public policy" (http://www.environmentaldefense.org, accessed September 2006). The goal of Scorecard is not the provision of scientifically justified or exact pollution data. Rather, the aims are to present data in such a way that it echoes the interpretation framework of lay actors, to stimulate lay actors to take an interest in local environmental quality and to provide a platform for discussion and action. Given the interest in the project, one can conclude that there is indeed a demand for comprehensive, easily accessible and understandable information on local environmental quality. A next logical step would of course be to involve citizen-consumers in data collection, in units and dimensions that are closely related to their daily practices and experiences, without losing control of validity. That would also challenge the data collection monopolies of state and companies and further enhance the countervailing powers of civil society.

Scorecard has been used not only by individual citizens to become informed, to send complaints and to take action. Journalists, NGOs, companies themselves as well as their rivals, shareholders, and local and national authorities are among the regular Web site visitors. The countervailing effects are then also working via a variety of mechanisms, and not in the last place through preventive anticipation by polluters. Sand (2002: 4–5) and others[12] relate the 'success story' of the TRI and the Scorecard site in the United States to one or more of the following characteristics: (i) the possibility of electronic communications via the Internet; (ii) reversal of the burden of proof for exemptions; (iii) enforcement by citizen suits; (iv) standardised data, facilitating comparison and 'performance benchmarking'; (v) reputational effects of competitive ranking on a firm's behaviour.

Webloggers

As I will elaborate further in Chapter 9, one of the major differences between the 'old' media (radio, television, newspapers) and the 'new' media (Internet and related services) is that the production of news is no longer monopolised but democratised. On the Internet, everybody with

---

[12] See, among others, van den Burg (2004); Karkkainen (2001); Fung and O'Rourke (2000).

access to and skills of using a computer can put news and information on the Internet (provided that no limitations are put in place by, for instance, undemocratic states; cf. Chapter 10). A most interesting new development in this is the so-called webloggers or bloggers: personal Web sites that contain up-to-date information that is usually not found in regular media. Currently, at least every politician of some stature has such a personal weblog, as well as many celebrities. But also numerous 'ordinary' citizens have designed their own Web page and keep these regularly updated with all kinds of information, personal views, comments on news events, and discussion contributions. According to Technorati, a weblog search service, in 2006 every day some one hundred thousand new weblogs were established.[13]

On numerous occasions, blogs proved to be an essential source of countervailing information against major information disturbances; or they proved a rich source of information where the large monopolies remained absent. Since September 2005, Google, the most used search machine on the Internet, also searches in weblogs (in addition to the normal searches in official Internet sites), giving evidence of the increasing importance of these blogs for information collection. Eyewitnesses publish their accounts in no time on the Internet, for instance, on occasions such as the major hurricane disaster in New Orleans in 2005, when governmental information proved inadequate. But, at the same time, blogs increasingly have been discovered by politicians, businesses[14] and the (old) media as useful sources for, or dissemination channels of, information. At the major old media conglomerates, journalists constantly surf through the Internet to find blogs with new information, making direct links between old and the new media.[15] Major

---

[13] *NRC Handelsblad*, December 13, 2006, p. 21. In March 2007, the Technorati Web site mentioned more than 175,000 new blogs every day. Bloggers update their blogs regularly to the amount of over 1.6 million posts per day, or more than eighteen updates a second (http://technorati.com).

[14] In 2007, less than 5 percent of all large and medium-sized companies had a corporate weblog. European companies were mostly lagging behind (2.5 percent), whereas U.S. companies were leading (with more than 14 percent). Also, Asia-Pacific companies were greater than the world average, with 5.5 percent of the companies having a corporate weblog (LEWIS Public Relations, 2007).

[15] For some examples of weblogs dealing with the environment, see http://dominionpaper.ca/weblog/environment/ and http://www.fumento.com/weblog/archives/environment/.

'news discoveries' such as those concerning the Monica Lewinsky affair first appeared on blogs before entering the national or global media. At the same time, newspapers are increasingly going digital and making their Web sites 'blog friendly', for instance, by enabling easy linking and including references to relevant blogs in their articles. The reason is that almost one-third of the visitors to the Web sites of major American newspapers come in via weblogs. Perhaps one of the best indications of the informational power of these personal weblogs is the fact that in 2005 the Chinese government severely restricted the building of and access to these blogs (see also Chapter 10). And, as will be elaborated in Chapter 7, companies are also discovering the dangers and potentials of weblogs in their encounters with environmental NGOs and critical citizen-consumers, designing and implementing 'weblog strategies'.

At the same time, these powerful blogs are of increasing interest for state surveillance agencies, especially following 9/11 and the war on terrorism. With the increasing possibilities of individual citizen-consumers to not only easily access all kind of information but also produce and disseminate information, power balances are shifting. And so are policies and practices of (state) monitoring and control. Surveillance of the Internet has become in most countries one of the fastest-growing activities of secret and intelligence services, police forces and other state organisations. Regularly, discussions emerge in OECD countries on the necessity of more strict governmental regulation of weblogs, or – alternatively – calls for a civilisation process of webloggers, for internal codes or for stronger webmaster control.

The question is of course whether this major influence of weblogs will be there to stay. How sustainable is the revolutionary change that weblogs are bringing at the moment to the monitoring and media landscape? With the skyrocketing of the number of weblogs, an overkill of information is within reach or already present, and the number of readers per weblog is drastically reducing. Personal weblogs have a low survival rate after three months; the large majority have become inactive by then. But, at the same time, there is a tendency for weblogs to become more focused and specific, addressing and being visited by smaller, more carefully targeted groups and organisations. One can expect that the major influence of the weblog phenomenon on (environmental) governance will remain, but that it will be a limited number of these more focused weblogs that account for this influence.

6. Conclusion

Environmental monitoring has changed dramatically since the 1970s and 1980s. Initially, there existed basically a state-sanctioned, clock-time collection of natural science data, with relatively simple technologies, reported selectively in hard copies, all within the borders of a nation-state. Now, we see a highly diverse set of instant (or timeless time) data-collection practices at various levels (from global to micro), by numerous actors, with increasingly advanced technologies (such as satellites and nanotechnology), and reported via the Internet. This has not only resulted in increasing amounts of environmental data and information becoming available. The kind of data and information, the ownership and availability, the circulation of data and the surveillance of human practices have all changed dramatically. In addition to outlining these changes in this chapter, we have balanced the idea that this would only result in increased surveillance of institutional actors over citizen-consumers, and noticed that it also means larger potentials for countersurveillance.

The consequences of these changing environmental monitoring arrangements are to be felt in environmental governance, in which state and nonstate actors position and relate themselves in various ways to environmental information in the politics, debates and struggles on environmental challenges. The next chapters especially focus on how states and state agencies, private and commercial firms and civil society organisations work with information in their environmental reform practices. Here, the focus switches from data and information collection – the main subject of this chapter – to the application and use of environmental information in environmental governance.

6 | Environmental state and information politics

1. Introduction

In Chapter 5, attention was focused on (innovations in) monitoring schemes and arrangements for obtaining environmental information. In this chapter, we begin by shifting our attention to governance through information: the informational strategies and activities to redirect social practices into more environmentally sound pathways. Although there are various actors involved in informational strategies and governance, this chapter will put state organisations central, whereas the following chapters pay more attention to economic actors and the private sector (Chapter 7), to environmental NGOs and civil society (Chapter 8) and to the media (Chapter 9). Thus, although this chapter provides crucial insights in the origin and start of informational governance and its dilemmas, to fully understand the dynamics and scope of informational governance, this chapter falls short. The subsequent chapters must also be read in order to grasp its full breadth and complexity.

In conventional analyses of the role of information in state environmental policy making, the key problems are identified as information gaps, transaction costs for obtaining adequate information and ownership of information (see Chapter 1). Most legal and economic scholars have focused on how to overcome these problems in strengthening sound environmental governance. Some of the monitoring changes analysed in Chapter 5 are of key importance in doing so: more data availability, lower data collection and processing costs and larger proliferation of data and information. But this chapter moves beyond such a straightforward idea that more, better and cheaper information is automatically resulting in improved environmental governance, by focusing on three main topics of environmental governance in the Information Age.

First, the idea that information is gaining in importance in state regulatory activities is brought together under various headings: right-to-know, information disclosure, informational regulation, data-driven regulation. We will start our analysis by explaining how these forms of informational governance *avant la lettre* emerged, matured and widened, in order to understand what is changing today in state-directed environmental reform arrangements and politics. The second main development in informational governance is the growing importance and role information and communication technologies play in state regulation and policy making, especially under the heading of e-governance. Finally, we will enter into various discussions on the consequences of informational governance with respect to democracy and uncertainties, and analyse potential – and actual – threats of emerging informational governance, related to symbolic governance, information closure and deregulatory programs.

2. Information politics as environmental regulation

One of the initial fields in environmental governance in which informational developments came together to make a significant transformation was in the right-to-know and information disclosure policies, which started in the 1960s and matured in the 1980s. These developments changed the processes of state environmental policy making and the role of nonstate actors.

Although most existing environmental policy-making systems for a long time favoured administrative and corporate secrecy, and thus monopolisation of environmental information in the hands of governmental authorities and corporations, initiatives on right-to-know and mandatory disclosure of environmental information have been growing since the early days of modern environmentalism. Because of the specificities of its political system, information disclosure programs and right-to-know legislation emerged first and are arguably still most advanced in the United States. There, most current environmental laws include right-to-know provisions, most notably, of course, the Emergency Planning and Community Right-to-Know Act of 1986.[1] This

---

[1] Community right-to-know ideas emerged in the 1960s in the United States in both federal and state laws, requiring public access to information that the government collected from corporations as well as government's own data. The U.S. Freedom of Information Act of 1966 is one of the earliest of its kind.

federal law was a congressional reaction to a number of incidents, most notably the 1984 Union Carbide incident in Bhopal, India, in which two thousand people were killed. But this law is part of a much broader set of activities, protests, pressures and claims in many countries in the 1970s and 1980s, brought together under the right-to-know denominator. In a significant number of OECD countries, this resulted in right-to-know legislation in the 1980s (even six had them installed in the 1970s). In general, European countries were much later with right-to-know legislation; and within Europe especially the Nordic countries and the Netherlands were clearly preceding countries such as the United Kingdom, France, Germany, other Mediterranean countries and Central and Eastern European states (cf. Sand, 2002). Overall, we can witness in time a development from right-to-know policies and activities to active information disclosure.

From right-to-know legislation to active information disclosure

The U.S. Emergency Planning and Community Right-to-Know Act of 1986 formed the start of a new wave of information disclosure programs and policies, which matured especially in the 1990s. These disclosure programs and policies of the 1990s and beyond – which will be further elaborated later in this chapter – differ from right-to-know policies and practices of the 1960s, 1970s and 1980s in their proactive publication of environmental information, without specific requests being made to do so. Before the 1990s, information was in principle accessible under the right-to-know legislation but often difficult to obtain in practice, as a time- (and resource-) consuming march through the (legal) institutions was often needed. More specifically, these earlier policies gave only fragmentary information and were not helpful for users to compare products systematically, to rank risks or to judge easily their own exposure. Second, the kind of environmental information is also different. The new disclosure systems collect information – also or even primarily – with the goal to inform the public, whereas the 1970s and 1980s information was just collected to inform the government. This makes the type, format, completeness and direct and instant access of today's environmental information very different from its predecessor. Third, and finally, the new disclosure systems hold polluters responsible for their action as well as for the correctness of the information provided; thus, information is not aggregated by industrial sector, geographical area or water body. Information is now increasingly specific on facilities, companies and products.

The first major mandatory environmental information disclosure program started in the late 1980s in the United States, the U.S. Toxic Release Inventory, TRI. During the 1990s, the U.S. Environmental Protection Agency (USEPA) substantially expanded the amount of chemicals for reporting and the number of firms that had to disclose their emission data. In the 1990s, there was a growing demand for and regulatory codification of right-to-know and information disclosure in the environmental policies of almost all Western industrialised countries. It is surprising to see that most of the national freedom of information acts have an environmental origin or background, and environmentalists can be found among the constituencies that have been pressing continuously for freedom of information.

The call for further public access to environmental information collected by polluters and state agencies got a new – now international – impulse following Principle 10 of the Rio declaration, agreed on at the United National Conference on Environment and Development (UNCED) in 1992. Within the European continent, this resulted in the Århus Convention in 1997,[2] whereas globally the Access Initiative and the Partnership for Principle 10 continuously kept access to and disclosure of information on the global political agenda. By 2000, forty-four countries had passed access-to-information legislation; twenty-four of them were OECD countries (World Bank, 2002).[3] Also, international organisations and institutions such as the World Trade Organisation and the World Bank were faced with increasing pressure – internally

---

[2] The Convention on Access to Information, Participation in Decision-Making and Access to Justice in Environmental Matters, was developed under auspices of the UN Economic Commission for Europe, was adopted in June 1998 and entered into force on October 30, 2001. In 2004, more than forty countries had signed the Convention and thirty-three had ratified the Convention, which also has provisions for non-ECE countries to accede. The Convention refers specifically to the active use of the Internet and electronic information disclosure in providing access to environmental information. For the EU, this Convention resulted in the revision of the EU directive 90/313 on access to environmental information into the much more powerful EU directive 2003/4/EC.

[3] In some countries, access to information even made it into their constitutions, for instance, in Thailand, Mexico, South Africa and Uganda (although the latter had not specified this constitutional right in a Freedom of Information Act as of 2002).

and externally – to disclose environmentally relevant information (see, for instance, Udall, 1998, on the changing World Bank policies in this respect). According to a World Resource Institute Report (Petkova et al., 2002), three main forces have triggered the growing worldwide demand for public access to environmental information: (i) the urgency and scope of environmental problems, which called for wide support and thus information sharing; (ii) the increase in activism in civil society, also outside the OECD countries; and (iii) the developments and spreading in information technology and communication means.

Although in most OECD countries access to environmental information has been institutionally safeguarded, the debate has not lost its urgency, for two reasons. First, many of the non-OECD countries have still not installed legal provisions and institutionalised practices for freedom of information and information disclosure, although changes in this regard can be noticed. Towards the end of the 1990s, these ideas of right-to-know and information disclosure began to spread beyond the OECD countries, towards industrialising countries in, among others, Southeast and East Asia (see Chapter 10). In addition, a number of international organisations and initiatives actively promote mandatory information disclosure and environmental reporting by private companies.[4] Second, the implementation of access to environmental information and the actual easy access of civil society to environmental information still seriously lags behind the legal codification. In a study on access to environmental information in a representative sample of nine countries,[5] it was found that the legal provisions and state of the environment reports were mostly in place, as well as the quality and accessibility of information on highly visible emergencies and air quality. But access to information on for instance water quality, industrial facility emissions and risks, and accidents at private facilities was generally poor (Petkova et al., 2002).

---

[4] For example, the 1992 CERES Principles of the Coalition of Environmentally Responsible Economies; the 1993 PERI guidelines of the Public Environmental Reporting Initiative; the various guidelines for reporting of the International Standard Organisation; the 1999 Sustainability Reporting Guidelines by the Global Reporting Initiative; the EU directive 2003/4/EC on public access to information. Other privately supported and initiated guidelines and initiatives are Goldman Sachs Best Practices; Malcolm Baldrige National Quality; Social Accountability 8000; The Business Council for Sustainable Development Corporate Government Principles; Global Sullivan Principles; and AA1000.

[5] Chile, Hungary, South Africa, Uganda, India, Indonesia, Mexico, Thailand and the United States.

Four main factors seem to determine the national performance on public access to environmental information in a country. First, the government capacity in staff, equipment, procedures, training and the like is a major factor in actual public access to environmental information. Sub-Saharan Africa gives evidence of poor state capacities, compared with, for instance, China and – to a lesser extent – Vietnam (cf. Chapter 10). Second, an active and capable civil society with strong NGOs contributes to effective use and good performance of public access. Countries such as China and Vietnam are clear cases with still weak civil societies, which to some extent reflect on the access to environmental information. Third, the position of the media, media attention and coverage of environment, and public scrutiny prove to be major factors in improving public access. And, finally, the international community, via provisions and requirements in multilateral environmental agreements and conventions, via official development assistance, via the operations of multilateral institutions and via international organisations, forms an important factor in assisting especially developing and industrialising countries to create information systems, provide staff training, publish state of the environment reports and support all kind of other conditions that facilitate public access to information. A special case is the European Union, where the accession process of ten new members from Central and Eastern Europe formed a major trigger for upward harmonisation in access to information.

Regulation by information

Initially, the right-to-know and mandatory disclosure legislation and ideas were not related to any ideas of creating incentives or new dynamics of environmental improvements and reform. The basic idea was simply related to ideas of democracy and transparency: citizens have a democratic right to access information and for that information to be disclosed publicly. It was only in the mid-1990s that information disclosure, right-to-know, and environmental reporting obligations were interpreted as having positive environmental governance effects. In the more legal and economic American and international literature, the influence of the wider availability of environmental information on environmental policy-making and regulatory processes have been brought together under the notion of informational regulation (e.g., Konar and Cohen, 1997; Tietenberg, 1998; Kleindorfer and Orts, 1999; Case, 2001; Dasgupta and Wheeler, 1996; Graham, 2002).

Informational regulation is then defined as rules and institutions requiring mandatory disclosure of information on environmental practices or performance of regulated entities to third parties. With informational regulation, standard conventional regulatory practices of states, such as standard setting, licensing and enforcement, are complemented or partly replaced by new informational dynamics in which other nonstate actors play a significant role: 'regulation by revelation' (Florini, 2003).

In the conventional regulatory system, the state relies on state-run, expert-led and natural science–based monitoring systems to see whether, where and when state regulation and enforcement needs to be intensified and policies have to be adapted. Under conditions of informational regulation new monitoring systems and mechanisms and new enforcement dynamics start to appear, and not only in the advanced industrialised settings.[6] Information disclosure to the public is then seen as a more effective enforcement mechanism than classical enforcement via the state. For instance, complaint systems (via letters, telephone lines and/or the Internet) and surveys are actively stimulated and organised by state authorities to set priorities in policy making, but also – when publicised widely – to assist conventional enforcement activities. According to Graham (2002: 10–11), mandatory disclosure strategies differ in three ways from conventional governmental environmental policies. First, these strategies influence environmental risk levels not through legislative or regulatory processes by the state but through nonstate public pressure. Second, the 'regulators' are not governments but the countless actions of numerous nonstate actors that are empowered by knowledge and information to change purchasing, investments, voting, collective actions and so on. Third, these systems extend beyond the reach of the government and beyond the national boundaries.

With the new information technologies this public disclosure, right-to-know and informational regulation has recently taken a major step forward. Especially since the 1998 UN/ECE Århus Convention,[7]

---

[6] See O'Rourke (2004), Phung Thuy Phuong and Mol (2004), Dasgupta and Wheeler (1996) and Brettell (2004) for examples of complaint systems and community-driven regulation in Vietnam and China, two still strongly state-dominated systems. See also Chapter 10.

[7] Under the UN ECE Århus Convention, a protocol has been negotiated in Kiev in 2003 on "Pollutant Release and Transfer Registers." In 2004, thirty-seven

the Rio Principle 10 and the Access Initiative,[8] OECD countries have sped up regulation on Pollutant Release and Transfer Registers (cf. Sand, 2002; Graham and Miller, 2005). Strongly supported by ICT developments, this has resulted in direct access to standardised, site-specific, up-to-date and user-friendly environmental emission and performance data of thousands of companies, among others, in the United States, Australia, the United Kingdom, Japan, Norway, the Netherlands and other countries in the EU.[9] Not only do these sites allow citizens to search their neighbourhood for polluting sources, to compare these facilities and neighbourhoods with others and to pressure companies for improvement, but they also moved journalists, polluters themselves, governmental authorities, political representatives, shareholders and others into environmental action, leading to environmental improvements. Fung and O'Rourke (2000) explain how the TRI mandatory disclosure program in the United States has triggered various social dynamics by different actors, all towards more environmental pressure being put on polluters and environmental state authorities: direct action by communities, pressures on regulatory agencies for enforcement, mass media attention, stock market reactions, new legislative initiatives by states in the United States and preemptive company actions. By 1998, the TRI mandatory disclosure program was credited with a 46 percent reduction in emissions of the listed chemicals over ten years (cf. Fung and O'Rourke, 2000; Graham and Miller, 2005).[10] Chapter 7 also will discuss how similar dynamics

parties had signed the Protocol, but none of them have yet ratified it. By the end of 2004, it had not yet entered into force.

[8] See the Web site http://www.pp10.org. In Chapter 8, we will further elaborate on the right-to-know and information access developments in the field of environmental governance, as these are especially meant to further involve citizens and civil society in environmental governance through information.

[9] See http://www.scorecard.org, launched by the Environmental Defense Fund in 1998; http://www.environment-agency.gov.uk, the U.K. government successor Web site of Friend of the Earth Factory Watch program, launched in the late 1990s; http://www.eper.cec.eu.int, launched in February 2004 by the European Union. Petkova et al. (2002) also provides data on countries that are taking steps towards pollutant release and transfer registers, which include Mexico and Slovakia.

[10] Of course, the data after 1998 is more complicated. The reductions in releases were particularly strong in the early years, it mostly happened via recycling rather than prevention, emissions of carcinogens decreased more than average, and the distribution varied greatly among industrial sectors, media (air, water, and land), and states (see Graham and Miller, 2005, for details). Note that this

work for product-specific disclosure programs, such as those related to labelling and product information.

It should not be too surprising that the landmark disclosure programs on pollutant releases have their own flaws and criticism. For instance, the TRI has been criticised for the fact that it neglected sources that significantly contribute to pollutants: mobile sources and small business (Graham, 2001). TRI also did not require factories to provide data on human exposure to chemical toxicity (only data on kilograms of emissions). Finally, the TRI reporting was based on a variety of estimating techniques and it was often more than a year old before the information was released. Several of these points were taken up by NGOs in launching the Web site Scorecard (see Chapter 5). A more fundamental point is that these Pollutant Release and Transfer Registers are still very much based on natural-science information. Communities, environmental NGOs, social scientists and even state authorities also organise nonnatural science–based monitoring and information disclosure programs, focusing on preferences, lay actor experiences and sensory information of citizens, consumers and more specific segments of society (cf. railway travellers, nature hikers, solar energy–producing households). With the disclosure of this type on nonnatural science environmental information, one could imagine further – and new – mobilisation of public opinion, pressure on state agencies and polluters and setting of new agendas for environmental reform.

Data-driven regulation

Recently, Daniel Esty (2001a; 2000b; 2004; Esty and Rushing, 2006) has further worked on the changes in state environmental regulation through information, towards what he labels data-driven policy making. It is especially data shortage and data quality that has hampered adequate and high-quality environmental policy making for a long time. Especially through the technological breakthroughs of computer

is all before the U.S. environmental movement got involved in TRI via its Scorecard Web site in 1998. Konar and Cohen (1997) reported clear effects on the stock price of TRI data disclosure. Hamilton (2005) reports on more recent empirical research in TRI. Edwards (2006) reports on case study research in the city of Cleveland (USA), where environmental NGOs played a major role in the environmental information structure and the modernisation of public administration towards a more citizen-centred mode of environmental governance.

and ICT the ability and capacity increased to spot and visualise environmental problems, assess their scope and seriousness, understand their effects and manage the information flows, all essential for identifying and designing better response strategies. But, according to Esty, perhaps the largest contribution to environmental governance comes from transparency, where information and data become widely available and disclose the laggards and polluters. This is valid within nation-states, as well as across countries.

The Environmental Sustainability Index, more recently relabelled as the Environmental Performance Index,[11] is a key example of what Esty means when he refers to the power of data in driving environmental reform. This index ranks countries in terms of their environmental performance on a large number of environmental indicators. As one of the founders and main contributors to this ESI, Esty (2004) illustrates how states are motivated by these disclosures by the example of the outrage of the government of Belgium when it found itself listed initially in position number seventy-nine. With this EPI program, environmental data collection, processing and disclosure is aggregated at the level of countries, whereas in the former disclosure programmes, it was usually at the level of firms (or municipalities; see Chapter 10). Taipale (2003) joins Esty in his analysis, by stressing that information technology has become an agent of change in recent state environmental policy making, moving environmental policy making away from its dominant command-and-control outlook towards more innovative market- and information-driven modes of governance.

Compared to the earlier notions of right-to-know, information disclosure and informational regulation, notions of data-driven policy making are much more technology driven and enabled. It is the new information technologies that enable and drive new modes of environmental governance, and not so much the activities of NGOs and civil society. Data-driven regulation is thus much more strongly connected to digitalisation, enhanced possibility of data processing

---

[11] This index has been developed by the Columbia University Centre for International Earth Science Information Network (CIESIN), the Yale Center of Environmental Law and Policy, Yale University, and the World Economic Forum's Global Leaders for Tomorrow Environment Task Force. See http://www.ciesin.columbia.edu/indicators/ESI. The Environmental Sustainability (or Performance) Index (ESI/ EPI) is a measure of overall progress towards environmental sustainability, developed for 133 countries. The ESI scores are based on a set of twenty-two core "indicators," each of which combines two to six variables for a total of sixty-seven underlying variables.

by computers, advanced technological monitoring (such as satellites) and new information-processing schemes. Next to the fact that these digital technologies have more data-processing capacity, speed and accessibility, they are also believed to change the modes of governance towards electronic governance.

3. E^2-governance

Every revolution in communication technology has given birth to utopias of conflictless democracy and governance, as well as of further rational planning efforts and state control. The electronic and cyberrevolutions of the 1990s initially let to equal high expectations of e-governance, e-administration and e-democracy, but soon also met pessimism and scepticism on the materialisation of these promises and the unwelcome side effects.

The notion of electronic governance or e-governance is widely discussed and entails a wide set of meanings, interpretations and practices in numerous areas of policy making and rule making.[12] In its widest connotation, it refers to all electronic activities by or on behalf of governmental authorities in the policy-making and implementation process. It then entails everything from electronic information dissemination on plans and measures to shifting part of the policy-making process itself from face-to-face to electronic exchanges and interactions. The practices of e-governance vary across the nation-states of the world. Table 6.1 provides evidence of the large divide in e-services on national governmental Web sites in different parts of the world. Using three sets of indicators for e-governance (related to information availability, service delivery and public access), West (2005) found that the countries with the most advanced e-governance systems were not only located in North America, Europe and East Asia, but they were definitely not located in Africa, Central Asia and the Middle East.[13]

---

[12] In the United States, the federal E-Government Act was approved in 2002, and before that there was an increased commitment by the government to Internet-based public participation strategies (cf. Zavestoski and Shulman, 2002). Also, in many other OECD countries, certain parts of electronic governance are getting more and more mainstream. Since 2004, the *Journal of E-Government* is being published – digitally, of course.

[13] In 2005, the top ten nations with e-governance, according to these indicators, were Taiwan, Singapore, the United States, Hong Kong, China, Canada, Germany, Australia, Ireland and the Vatican. This was only slightly different in 2004.

Table 6.1. *Percentage of national government Web sites offering e-services, for different world regions, 2001–2005 (West, 2005)*

	2001	2002	2003	2004	2005
North America	28%	41%	45%	53%	56%
Pacific	19%	14%	37%	43%	24%
Asia	12%	26%	26%	30%	38%
Middle East	10%	15%	24%	19%	13%
Western Europe	9%	10%	17%	29%	20%
Eastern Europe	–	2%	6%	8%	4%
Central America	4%	4%	95%	17%	15%
Latin America	3%	75%	14%	10%	19%
Russia/Central Asia	2%	1%	1%	2%	3%
Africa	2%	2%	5%	8%	7%

The goals of new electronic governance and rule-making technologies are generally considered to be fourfold (Coglianese, 2004): (i) the increase of democratic legitimacy, (ii) the improvement of regulatory decisions, (iii) a decrease in administrative costs and (iv) a better regulatory compliance. In its most positive assessments, e-governance shows improved agency efficiency going together with cost effectiveness, transparency, responsiveness and accountability. Arguing in this positive and optimistic line some claim that Internet and new information technologies have a potential for "distributed, reflexive, transparent, information-rich, asynchronous, widespread, low cost, meaningful and transformative participation in timely decision-making" (Shulman, 2004: 8), thus enhancing democratisation (e.g., UNESCO, 2005). Others, however, fear that the sheer mass of electronic information and participation will overwhelm the decision-making process and thus delay and even frustrate public authorities with their limited capacities. In addition, the digital democracy can reproduce or even enhance inequalities through the digital divide, where major groups and communities remain outside the informational flows (cf. Wilhelm, 2000).[14] These general claims are also made with respect

---

[14] In discussing these digital-induced inequalities in one of the most digitalised nation of the world – the United States – Wilhelm (2004: 73) notes that 112 million Americans are not online, 90 million are defined as low-literate, 53 million have some level of disability, and 25 million residents do not speak English at home. Online government portals at twelfth-grade levels, only in English, and without being disability-accessible marginalise millions of

to environmental e-governance (cf. Zavestoski and Shulman, 2002) or e^2-governance.[15]

Perhaps one of the more extensive and interesting explorative investigations into the potentials and innovations of e-governance for environmental policy is done by the eRulemaking Research Group in the
United States (cf. Shulman, 2004).[16] In a number of workshops with
policy makers and stakeholders in policy-making processes, the influence of the Internet on environmental rule making was assessed. From
these workshops, it proved that there is still little agreement and large
uncertainty about how policy- and decision-making procedures and
outcomes will change following the growing development of electronic
governance. The claims of expanded public access and growing influence of civil society and citizens have not really been fulfilled yet. Lack
of harmonisation and centralisation among governmental agencies in
e-governance, user-unfriendly systems and inequalities because significant groups have difficulties in accessing Internet and commenting elec-
tronically are among some of the downsides of e^2-governance witnessed
by citizens. By the same token, environmental policy makers and rule-
makers are confronted with questions of usability of large quantities
and poor qualities of electronic submissions and have to learn how to
deal with such comments; NGOs face the inflation of numerous duplications of similar comments submitted to rule makers, especially in
those situations in which electronic commenting is also facilitated by
environmental NGOs to increase membership.[17] Consequently, much
governmental effort is recently put into procedures and technologies to
organise and summarise the contents of large public comment
databases, experiences also witnessed in other European countries.

U.S. citizens from benefits and democratic processes they are legally entitled to.
See more general analyses of (cyber)democratic governance in the Information
Age in Kamarck and Nye (1999).

[15] Note that much of the explorations in this section and chapter refer to
e-government rather than to e-governance. Thus, discussions here will not
include questions of the governance of cyberspace information at large (for
that, see Kobrin, 2001).

[16] This research group consists of scholars from various disciplines and a number
of different U.S. universities. It developed out of a workshop held at the
Kennedy School of Government in 2003. See also their project Web site
http://erulemking.ucsur.pitt.edu/.

[17] Shulman (2004) reports on U.S. Environmental Protection Agency notices on
mercury emissions, which resulted in 680,000 e-mailed public comments. After
a labour-intensive manual sorting, there proved to be forty-two hundred
distinctive comments.

At the same time, structured input forms, rebuttal comments (and eventually online dialogue), open modelling to assist in understanding technical information and other innovations are suggested to facilitate and further participation in e-governance.

As Tumber (2001) rightly remarks, electronic governance also can have major consequences for the role of the conventional – up until now powerful – media and forms of mediation in governance. It changes or even replaces the government-media-public system and makes governments and other agencies and organisations, such as NGOs and other private actors, less dependent on the media. Even such powerful media as television are challenged by the Internet, through which information production and transmission is democratised. In Chapter 9, I further explore how the media is involved in environmental governance and how its role and position changes under conditions of the information revolution.

Overall, the potential change that e-governance brings to regulating the environment can be large. But whether that really materialises and what the actual nature of this change is remains uncertain. Increased transparency, participation, accountability and better decisions can be outcomes, as well as procedural (rather than substantive) legitimation, information overkill (both for citizens and for governing agencies confronted with comments), further inequalities in influencing rule making through a digital divide, growing litigation and the undermining of traditional forms of participation without providing an adequate digitalised substitute (cf. Schlosberg and Dryzek, 2002; Wong and Welch, 2004). But digital environmental policy making and rule making also might basically digitalise the existing paper-based policy-making process, copying the strength and weaknesses of existing – nondigitalised – policy processes, as Shulman (2004) seems to conclude: "The Internet Still Might (but Probably Won't) Change Everything." This will be especially the case when the rationale for e-government remains in cost-savings and other familiar efficiency matters (or what Chadwick and May [2003] label the managerial model of e-government), rather than focusing on the potential gains towards more participatory democracy and higher levels of trust, transparency and legitimacy.

4. The search for information quality

What the developments in informational governance and e^2-governance point at is the increasing role of information, informational processes, and information technologies in conventional policy processes, and with that the transformation of these conventional processes. With the growing role and 'power' of information in policy-making, rule-making, and decision-making processes new questions emerge and old questions are being put in a different light. Here we will focus on the ways in which policy makers deal with new questions of uncertainties and informational controversies.

Uncertainties have always been present in environmental governance, whether they were uncertainties on the sources of environmental pollution, the quantity of natural resource reserves, the (short- and long-term) environmental and health effects of practices and substances and/or the effectiveness of policy measures and strategies to combat these. A wide literature has explored how these conventional uncertainties transformed and – according to some scholars – radicalised under conditions of late, reflexive or global modernity (Wynne, 1992; 1996; Jasanoff, 1996; see Chapters 2 and 3). Decreasing authority of science; increasing amount, speed and distances of information generation, transfer and availability; and the transformation of state authorities in policy processes are crucial factors in failures at closing controversies around environmental risks and in continuations of uncertainties. Once information becomes of crucial importance in environmental governance, these uncertainties are more than just unwelcome side effects of an unquestioned governance model. States, public policy makers and scientists cannot but address these new challenges and in doing so they develop various strategies, new arrangements and suggestions. Here we will focus on two – not necessarily independent or mutually excluding – ways in which policy makers deal with issues of uncertainties and informational controversies: better information and more participation.

The call for information quality: the U.S. Data Quality Act

One typical reaction, especially strongly felt in Anglo-American policy cultures, is the quest for further certification of the quality of information used in environmental governance. The increasing amount of information circling around governmental interventions calls for procedures to distinguish useful from useless information, especially in times when information is becoming so powerful. The U.S. Data Quality Act (DQA) is a typical example of this type of reaction.

In 2001, the U.S. Congress – strongly pushed by industry (Noe et al., 2003: 10226; Herrick, 2004: 420; the Water Resources Research Institute, 2004) – passed the Data Quality Act,[18] requiring agencies to establish procedures to ensure and maximise the quality, objectivity, utility and integrity of the information they disseminate. The Office of Management and Budget (OMB; and especially the Office of Information and Regulatory Affairs at the OMB) was provided with the task to interpret and operationalise this one-sentence legislative requirement. No legislative history indicated what it was that Congress actually meant. The OMB interpreted the background of this Act[19] to be located in the Information Age, where information has become a vast resource of power and where governments rely increasingly on information dissemination to accomplish its goals. "Regulation by information is becoming the norm" (Noe et al., 2003: 10224). Reliability and quality of information is then essential for the public. OMB has developed government-wide and agency-specific guidelines for governmental agencies to fulfil the requirements of this Act. Although the Act covers all federal agencies, the focal point in much of the discussion has been on environmental and health issues and thus on the U.S. EPA, also because EPA regulations and policies are often based on emerging science and can have a major impact on vested interests.[20] For the U.S. EPA, a fifty-four-page guideline has been developed, defining standards of objectivity, quality, integrity and

---

[18] Section 515(a) of the U.S. Treasury and General Government Appropriations Act for Fiscal Year 2001 (Public Law 106–554). The origin of the Act was in a political dispute over air pollution, when the U.S. EPA proposed to tighten national ambient air quality standards for fine particulates and opponents felt unable to assess and review some of the supporting scientific data. The industry-sponsored Center for Regulatory Effectiveness was the main lobby for this Act.

[19] OMB prefers to refer to this Act as the Information Quality Act. It covers more than just data but also regulatory, statistical, research, financial, risk assessment and other governmental information. It also includes third-party information 'initiated' or 'sponsored' by governmental agencies. Distinctions are made between ordinary information and influential information, with different quality regimes. A major discussion is whether the Act applies to rule making as well (cf. Noe et al., 2003).

[20] In the Fiscal Year 2003, EPA received thirteen requests for corrections based on the DQA, only to be surpassed by the Department of Transportation with thirty-eight petitions (Herrick, 2004). For corrections of information filed with the EPA, see http://www.epa.gov/oei/qualityguidelines/af_req_correction_%20sub.htm.

utility of information.[21] Parties affected can – based on this Act – petition against governmental information that is not according to the guidelines. The Act requires "government agencies to ensure the quality of data they use when issuing new rules, regulations and studies. For the first time, anyone will be able to challenge the data used in formulating government regulation, instead of just challenging the rules themselves" (Horvath, 2002).

Although the principle of improving information quality is hardly questioned, the consequences of the particular DQA and the guidelines installed by the OMB are. It is expected by environmental NGOs[22] that regulated private sectors will use the Act as a weapon to delay the ability of agencies to put new safeguards in place and address environmental and health issues. It also might be used to suppress information dissemination that is so essential to inform the public in a democratic process. This is aggravated by the fact that petitions and complaints against agencies are dealt with secretly, behind closed doors, by the agency, the affected party and the OMB, without informing the public. In addition, environmental advocates wonder how this DQA relates to the precautionary principle: as conservatives and private sector interests usually interpret the application of the precautionary principle as a cover for imperfect information and bad analysis, would the DQA (further) obstruct the application of the precautionary principle? Finally, there is a risk that this DQA will inhibit information transparency and chill dissemination, as agencies may find it too troublesome to fulfill the DQA guidelines and rather chose for not publicising information than going through a troublesome process of examining data quality. This would weaken the role of public discourse in policy formulation and regulatory oversight (Herrick, 2004). And, indeed, OMB Watch, a U.S.-based nongovernmental organisation dedicated to promoting government accountability and citizen participation in policy decision making, has recently found several cases of strategic use of the DQA

---

[21] EPA, *Guidelines for Ensuring and Maximizing the Quality, Objectivity, Utility and Integrity of Information Disseminated by the Environmental Protection Agency,* can be found at http://www.epa.gov/oei/qualityguidelines/. Of additional interest are the guidelines that the EPA has produced for assessing information from external sources to be used by EPA: *Assessment Factors for Evaluating the Quality of Information from External Sources* (http://www.epa.gov/oei/qualityguidelines/af_assessdraft.pdf).

[22] See the comments of the U.S. environmental NGO Natural Resources Defense Council on the DQA in Noe et al. (2003).

by companies to challenge EPA choices and policies.[23] Via this Act, the growing regulation by information of many agencies is counteracted by industry and business via regulation of information. As such it should be interpreted as an attempt to control and set back informational governance, by bringing it under stringent (state and elite) control.

Behind this DQA is a rather narrow and positivist idea of objectivity and validity of information. Quality of information seems to lie in the internal qualities of the information, irrespective of the context and the circumstance of its application, or what Herrick (2004) calls narrative coherence, which is a function of, among others, contextual and procedural relevance, methodological rigor and transparency. It neglects that, especially in the Information Age, "there is no unique way to construct an argument: data and evidence can be selected in a wide variety of ways from available information [...] there is nothing intrinsically reprehensible in selecting a particular combination of data, facts, values, and analytical methods that seems to be the most appropriate to convince people who have to carry out (or support a) decision" (Majone, 1989). In that sense, the DQA is in line with (or tries to get back to) the conventional idea of science, one that belonged to an era of simple modernity, whereas there is growing consensus among social scientists that science and information have lost their undisputable position and reputation. But such an emphasis on positivist science and the objectivity of data and information also can be found in Esty's project of data-driven regulation. Esty (2001a; 2001b; 2004) has – among other things – a similar emphasis or more and better data and information to improve the quality of regulation and decision making.

However, what distinguishes the Data Quality Act and its proponents from the data-driven regulation analysis is that the former falls in a wider development in particularly the United States, characterised by calls for sound science in a green backlash (see also McCright and Dunlap, 2000; 2003). The use of climate science for the official U.S. position on international climate negotiations have been widely studied. Another less well-known example is the EPA-sponsored study on *Science at EPA: Information in the Regulatory process* (Powell, 1999).

---

[23] See the Web site of OMB Watch (http://www.ombwatch.org) and especially the part on EPA policies and decisions. Interestingly, OMB Watch also keeps track of the consequences of 9/11 for (environmental) information disclosure (http://www.ombwatch.org/article/articleview/213/1/1#EPA).

The book provides an interesting, detailed and critical analysis of the way in which scientific information is used and applied in regulatory processes, calling for various changes and budget increases. But the study has been used, for instance, by Singer (2000), to vehemently criticise and discredit the EPA for its use of science and information and to call for taking scientific research and budget out of the U.S. EPA. Indeed, this is another example of the ways in which data, information and knowledge (and with that science) have become central environmental battlefields with the emergence of informational governance.

5. Participation, trust and transparency

A second main line of dealing with environmental controversies and uncertainties in an era marked by information flows and disenchantment with science is through participation.

Participation has been part of environmental governance – and called for by policy makers and stakeholders – ever since industrialised societies faced major disturbances of their sustenance base. For a long time, however, public participation followed – aside from normative commitments to democratising policy making – the 'information deficit model'. Although participation was seen as essential for effective policy measures, the public was believed to be ill-equipped to take actions and decisions in the interest of the environment. In persuasive logic, the objective of participation is to provide further information to the wider public to make their actions fitting the implementation of governmental measures and policies. Recently, with growing complexities and interdependencies, with new roles and positions of environmental states and environmental science/scientists, and with the emergence of uncertainties and unstructured problems on the political agenda, the call for and practices of participation in any program of environmental governance have been changed dramatically. In both situations, the old and the new, there remains a straight line between information and participation. But although in the 'information deficit model' participation is only possible through more information from one side (science and government), under current conditions participation is needed to close information controversies coming from different sides. Whereas the DQA aims to close information battles primarily by distinguishing sound (scientific) information from poor-quality information, a participative strategy makes no such distinction and treats different sources of information as in principal equally relevant. Not information quality but 'deliberative democracy' (e.g., Pellizzoni, 2003) should then close informational controversies.

Under the new conditions of late or reflexive modernity, new participative institutions are seen as a necessary condition in order to deal with the large quantity and variety of information in public policies. In a kind of Habermasian tradition, decisions made through a (local) participative process are regarded as intrinsically more 'good' and more right than others (Macnaghten and Jacobs, 1997; Sagoff, 1988). These ideas have formed the basis for various experiments with public participation in environmental governance processes, including (but not limited to) consensus conferences, citizens' juries, round-tables, and focus groups. The implications of the various forms, modes and experiments of participation for the policy process are ambiguous, perhaps even more in an era of digitalised governance. Noveck (2003) details the informational problems on the Internet (disinformation, information overflow) and the challenges of enacting accountability, transparency and deliberative governance through the hyperconnectivity of cyberspace. But also outside cyberspace, deliberative democracy has its challenges. Behind the seemingly practical issues of sometimes huge resource spendings through participative processes, questions on who should be involved and how should participatory processes be integrated into formal state policy-making processes lie deeper questions of representation and outcomes (Bloomfield et al., 2001; O'Neill, 2001; Owens, 2000). This raises the critical issue as to "whether there can be any direct correlation between a more participative democracy and environmental protection... there are no guarantees that procedural democracy will produce substantive environmental benefits if there are competing views of what the environment should be like and what it is valuable for" (Davies, 2001: 80). The enthusiasm that has been shown for new modes of and institutional arrangements for participation in environmental politics and policies is often tempered by the recognition of the complex and fundamental questions that these processes raise: how they are to be integrated into processes of policy formation and how to conclude among a range of stakeholders, experts and the public, whose views are all considered valid but opposite.

Although participation serves or means to serve democratic goals, Jasanoff (1996) questions the often-celebrated relationship between participation and solving environmental controversies and uncertainties. It is not so that the most open decision with most opportunities for participation necessarily leads to the greatest transparency with maximum public access and accountability, and the best solution for controversies. Open and highly participatory decision-making systems do much better at producing large amounts of information and endless duelling between experts of different kinds than at ending disagreements and solving controversies on uncertainties. It is especially the increasing complexity and scientific-technical nature of environmental information that puts participation and openness in conflict with solving disagreements. Participation does not resolve questions such as which problems are most salient; whose knowledge is most believable; and who and where is the authority located to end debates and controversies.

Trust and transparency

So, in the end, with the increasing flows of environmental information and the changes in the old institutions that were believed to be able to distinguish right from wrong and truth from fiction, trust has become of key importance. New forms of informational governance relate truth claims with trust in the bringer of the claim. In contrast to the DQA, the inherent quality of information is not seen as decisive, but, rather, the quality and trustfulness of the information provider and generation. Carolan and Bell (2003) are correct in suggesting that truth has thus become a social relation and as such an essential part of the solution for dealing with informational uncertainties. Governance is not to make truth less social but make it more (explicitly) social. It is only via trust that truth and fiction can be and are separated.

But trust has changed in the Information Age. Trust in individual persons and leaders has become less and less relevant and is increasingly replaced by trust in institutions, in procedures, and in social/experts systems. The various opinion polls (such as those by the *Eurobarometer*)[24] that rank reliability of institutions in their informational role in environmental governance provide clear indications of where trustful and reliable – and thus powerful – institutions in informational governance are located. The constant high credibility and reliability of NGOs

[24] Cf. European Commission (2005), *The Attitudes of European Citizens towards Environment*, Eurobarometer No. 217 (carried out by TNS Opinion and Social), Brussels: DG Environment EC. The investigations among approximately twenty-five thousand citizens in twenty-five member states were done between 27 October and 29 November 2004.

vis-à-vis, for instance, politicians and business is a relevant factor in the new power relations in informational governance of the environment. But trust is not a static concept and needs to be built and (re)gained continuously, as Greenpeace and Shell witnessed following the Brent Spar controversy in the mid-1990s.

Procedures also gain trust. In informational governance, especially when global Internet-based flows of information are at stake, transparency and accountability are key trust-building mechanisms. This all is not to say that the content of information plays no longer any role or that all information is of equal value. Conventional criteria of coherence, consistency, objectivity and verifiability are still crucial first steps in assessing information quality, and conventional mechanisms as peer reviews still do their work as the controversies surrounding Lomborg's *The Skeptical Environmentalist* may illustrate. Argumentative debates and thus discursive democracy remain crucial around controversies and uncertainties, but they will not close the debates. In an era in which information is overwhelming, multidirectional, never consensual and a key resource in environmental governance, the combination of information quality with trustworthy actors and transparent and accountable information-handling processes has become crucial.

Hence, we should not be surprised to witness the large efforts various state and nonstate organisations put in building and gaining trust and in developing procedures and strategies for transparency, accountability, verification and reputation. And we should not be surprised how valuable reputations, trustworthy brands and reliable organisations have become, as compactly brought together under the denominator of 'reputational capital' for producers and 'legimatory capital' for NGOs. And as reputations are difficult to get but easy to lose, a widening and shift in the environmental battleground has taken place: from only formal politics to domains of trust, reputation, legitimacy, reliability, transparency and information control. Hence, these are the dimensions of information politics, which have complemented and partly replaced the conventional dimensions of laws, regulations and enforcement.

6. Regressive information politics?

The growing role and importance of information in environmental governance also involves various risks and potential regressions in environmental reform. Informational governance ideas and programs are regularly articulated and emphasised as an alternative to (conventional) state regulation, in the form of neoliberal, deregulatory programs. Although this does happen, I do not think this is a specific threat to informational governance, but, rather, a relatively independent development that all environmental programs and strategies potentially face.

Informational governance or regulation should not be put on an equal footing, or in direct causality, with voluntary measures or deregulation, so strongly welcomed by conservative business agendas. Although several informational innovations have been brought forward with the idea of regulatory retrenchment, and some hope that informational governance comes in place of compulsory legislative requirements, reality is different. The DQA example, for instance, shows the desperate need felt by business to further regulate informational governance in order to remove the environmental teeth of informational governance. Required transparency and mandatory disclosure do not just simply result in regulatory relief, nor do they just add to the regulatory 'burden' in environmental politics. Informational governance cannot be put into a simple one-dimensional regulation-deregulation scheme. As Mary Graham (2002) has powerfully illustrated, informational governance and mandatory disclosure are constantly threatened with arguments on protecting trade secrets, minimising regulatory burdens for companies, and guarding national security, especially after 9/11. Sometimes this results in calls for further regulation of informational governance measures, sometimes in calls for deregulation. Thus, although informational regulation is sometimes supported with arguments for further deregulation, I do not think that the two are very tightly linked, which discharges me from specifically entering into the potential threats of informational governance for deregulatory and neoliberal environmental politics.

In addition to this danger of deregulatory programs, I will discuss two other environmental reform threats of informational governance: symbolic governance and information closure.

Symbolic governance

The first references to symbolic politics date back from the 1960s (e.g., Edelman, 1964) and, since then, the idea of symbolic politics has received constant attention. In most of these studies, symbolic politics are interpreted in a rather problematic way, referring to politics or policy measures that are introduced only for rhetorical matters, often with the aim of satisfying certain parts of the electorate without having any serious intention for real implementation. Also within the field of environmental politics and governance, symbolic policies have been identified frequently among scholars (e.g., DeSombre, 2000; Matten, 2003), often sharply criticising the symbolic nature of measures and policies in which stringent implementation and enforcement are necessary.

Recently, however, symbolic environmental politics has been interpreted differently, by including the more beneficial or at least ambivalent sides of symbolic politics. Especially through the work of Hansjürgens (2000) on symbolic environmental politics, more emphasis has been put on symbols and on the way symbolic elements and even symbolic measures in politics are essential in communicating, motivating and activating environmental reform paths. Although symbols and symbolic politics might at first glance prove ineffective in directly and immediately addressing problems through concrete and adequate actions, there is increasing acknowledgement that they do play a critical role as an orientation and motivation around which ideas, perceptions and actions converge. Especially in situations with high transparency, visibility and significant public commitment symbolic politics and regulations do influence industries and consumers towards behavioural change. Under conditions of widespread environmental awareness and increasing information flows, interpreting symbolic politics just as an easy way to mislead the electorate and public does no longer stand. The media and other information and communication channels, in all their variety, play a crucial role in transferring, articulating and empowering these symbolic measures into actual actions and behavioural change. In that sense, interpreting symbolic environmental politics only in the conventional – critical and rhetorical – way has to be seen as a typical product of the pre–Information Age, in which transparency, visibility, accountability, information disclosure and time-space compression of information flows were largely absent or insignificant.

One step further would then be to interpret and apply symbols in environmental politics in a much more positive way. Signs, symbols and tokens (that is, information codified in a specific way) address and mobilise actors in a way that is often not possible via more conventional information channels, media and forms. Green colours, ecolabels, the WWF Panda logo and many other signs and symbols communicate environmental qualities but at the same time emphasise and articulate the essence and importance of environmental rationalities. And as such they govern actions: of consumers, industries, utility providers, citizens, local governments and so on. This raises a whole set of questions (on verification, symbol management, marketing of symbols, competition between symbols, actual behavioural changes, etc.), which we will address later in this book.

Of course, that is not to say that conventional symbolic politics no longer exist: governments, polluters and organisations try to mislead the public by avoiding serious environmental reform through symbolic actions. The Information Age literature shows us that there is a major chance that such symbolic actions may return as a boomerang on the source; either through demystifying these nonmeasures through transparency and public scrutiny or through self-fulfilling prophecy, when various actors start behaving in line with these symbolic measures, being afraid of damaging their reputational or legitimatory capital if they ignore them.

Information closure? Environmental governance and homeland security (post-9/11)

The growing importance of information in environmental governance makes environmental governance also vulnerable to information closure tendencies. Although from the 1960s onwards we have witnessed an overall tendency of further information disclosure, this is by no means a law written in stone, or an unfolding evolutionary process. The most challenging recent case is perhaps the aftermath of 9/11.

The terrorist attacks in New York (2001), Madrid (2004) and London (2005), among others, undoubtedly have had an impact on the debates on information disclosure. Immediately after 9/11, under significant time pressure and without sophisticated analyses, U.S. EPA officials made decisions on restricting access to sensitive information (cf. Graham, 2002: 6). The first response of EPA after 9/11 was to remove the Risk Management Plans from the EPA Web site, the most controversial section being the Offsite Consequence Analysis. In the aftermath of 9/11, a discussion evolved about the merits and dangers of providing potentially sensitive information on environmental risks to the general public. On the one hand, environmental interest groups and advocates of disclosure, such as the U.S.-based OMB Watch, stressed that the availability of information should not be comprised solely because terrorists might use the information. Their argument was that the safety of chemical facilities should be improved instead. Industries and parts of the state, on the other hand, argued that disclosure of information could be to the benefit of terrorists and should therefore be limited. No doubt this homeland security argument also was strategically used by industry in their constant struggle to limit environmental information disclosures. And it appeared that during the first years after 9/11 the media, especially in the United States, lost much of their critical attitude towards the government and governmental information systems, lowering media pressure for disclosure and accepting limitations on freedom of information.

In U.S. environmental politics, such struggles always have a strong legal component. Attempts to build a coherent legal framework that would make chemical industries less vulnerable for terrorist attacks – such as the proposed Corzine Bill – were dismissed as being 'stalinesque' and a 'jihad against chemical companies'.[25] The only two laws that were actually implemented to improve the safety of chemical facilities – the Public Health Security and Bioterrorism Preparedness and Response Act and the Maritime Transportation Security Act – cover only a fraction of the hazardous chemical facilities in the United States.

Various researchers studied the potential usability of the disclosed information for terrorists' purposes. In a research sponsored by the National Defense Research Institute, Baker et al. (2004) applied a supply-demand approach to information. In assessing whether or not terrorists would be able to acquire the information required, it was concluded that the 'flexibility' of terrorists (vis-à-vis the state) was to their advantage, adding that they have various means of acquiring information for their 'missions' including direct observation. On the supply side, the information currently available through federal Web sites was not considered crucial because the information was often spread across various state and nonstate Web sites or because the available information on federal Web sites was simply not relevant for terrorists. Fewer than 1 percent of the 629 federal databases were believed to contain sensitive information, and the accessibility of these databases was often already limited. Furthermore, closing the federal databases was believed to have a limited effect because much of the information

---

[25] See http://www.commoncause.org for more information.

was already mirrored on other – privately run – Web sites. Others argued that the truly relevant information, for both concerned citizens and terrorists, is the very detailed information about local facilities. So, although the availability of general information, such as aggregate chemical usage, is not at risk, one could – according to some – consider limiting the availability of detailed information to, for example, local community leaders or carefully screened individuals (see Cohen, 2002).

The direct consequences of 9/11 for environmental disclosure schemes, such as the TRI, appear to have been limited. After a brief period of heated debate not much has changed. In July 2005, updated information about the Risk Management Plan was suddenly released again, after OMB Watch filed a complaint in court.[26] Strangely enough, almost at the same time the national government of the Netherlands decided to limit information on their Web-based industrial risk information system exactly because of terrorism and homeland security.[27]

The long-term consequences are uncertain, however. Although the threat of terrorism may have had limited direct effects on information disclosure until now, it is used strategically as an argument for opponents of disclosure and debates have restarted on the costs and benefits of environmental disclosures (cf. Beierle, 2004). We also witness that the 'war on terrorism' revitalises the role of the national government in a wide range of policy fields, vis-à-vis nonstate actors. Concerning disclosure, we are likely to experience the (re-)involvement of national governments in balancing the need for information with the threat of terrorism. At the same time, and in contrast, we see that by the year 2005 in the United States, the media seem to have regained some of their critical stance against governmental information systems and do no longer easily accept information closures. The hurricane Katrina (2005) – and the strongly criticised governmental misinformation and lack of information – was instrumental in reactivating calls for information disclosure. Again, this illustrates the importance of the environment in debates on information (dis)closure. In moving from

---

[26] See http://www.ombwatch.org/article/articleview/2915/1/97?TopicID=1.

[27] The Ministry of Internal Affairs had a major conflict on this with the twelve provinces, as the latter refused to limit information disclosure on industrial and environmental risks and dangers to citizens. Finally, it was decided by the government on 9 September 2005 to withhold information from 1 January 2006 onwards on the environmental risks of facilities, but not on the facilities themselves, arguing the threats of terrorist attacks in the Netherlands make it that "safety is now more important than unconditional access to public information."

conventional ideas of one-dimensional security along lines of military and terrorist endangerments, to multidimensional definitions of security that include notions of environmental and ecological threats and risk, the opposition against information disclosure by using (national) security arguments is put in a different light (cf. Barnett, 2001; Pirages and Cousins, 2005).[28]

The informational state?

Ideas of regressive information politics are further developed and generalised by Sandra Braman (2006) into the thesis of a change of state: the (bureaucratic) welfare state of the twentieth century is replaced by the informational state in the new millennium. And – according to the author – this is not a trend to be welcomed or celebrated. This informational state uses information policy to exercise power, at the detriment of – among others – democracy, free access to information, information rights, evidenced-based policy making, and participation of citizen-consumers (Braman, 314ff.). In a lengthy and powerful analysis, Braman argues that the growing dominance of information in policies and politics and the way information is captured and monopolised by especially the state (and capital) leads at least in the United States to a surveillance state that threatens individual rights of citizens. In that sense, her thesis is in line with the critics of the Information Society such as Schiller and Lyon (cf. Chapter 2) and falls short because of a similar massive determinism by hardly allowing any tendencies of countersurveillance, transparency or civil society empowerments.

7. Conclusion: continuities and discontinuities

This chapter has illustrated and debated the information-informed innovations in the role of the state in contemporary environmental

---

[28] In April 2007, the Security Council of the United Nations discussed environmental security, endangered by climate change, for the first time. Despite opposition from the United States, Russia and China, which made clear that they did not see climate change as an appropriate subject for the security council, the United Kingdom pushed for this subject and in the end did not meet any veto. But the absence of most of the five permanent and ten rotating members was telling. At the same time, eleven former generals of the U.S. military issued a sixty-three-page report calling on the Bush administration to do more to counter climate change, warning that otherwise there could be "significant national security challenges" to the United States.

governance. With informational governance, the environmental state becomes engaged in new modes and dynamics of governance, with different resources, debates and politics. These informational changes partly parallel the wider discourse on shifts in governance, as reflected on in Chapters 1 and 3: more flexibility, more actors on stage (as we will see in Chapter 7 and 8), less hierarchical, a more dispersed locus of authority. But with *informational* governance the digitalisation, the centrality of informational quality, the importance of trust and transparency, and the politics and struggles on information move to the fore. In that sense, the notion of informational governance add to our insights on the locus of debates, struggles and politics that the environmental state is involved in nowadays.

Among all these innovations, we should be careful not to lose sight of the continuities in the role of the (environmental) state in governance. Governance innovations in mandatory disclosures and e-governance clearly illustrate the close relationship between conventional regulation and informational governance. In most examples and practices of informational governance, government and state authorities play still a major role, for instance, as the authority that requires the collection and disclosure of information or the organisation that mediates in or helps closing informational controversies. Also, in another sense, it is business as usual in informational governance: sometimes informational governance is supported by the powers that be as it is in their (e.g., deregulatory) interest; sometimes it is condemned by these very same powers and met with calls for further regulation of informational governance or complete deregulation. Following these controversies and conflicting interests on informational governance, disclosure is never full disclosure: some sources, some substances, some risks, and some societal activities, products and services fall outside disclosure and informational programs. In conclusion, there is significant continuity from conventional policy making into innovative informational governance.

As with conventional environmental policy making, the organisation and design of informational governance is thought to be crucial for its success. Access to Internet, understandability for a wide audience, trust in the medium and the messenger and interest within the wider public or their representative organisations are mentioned as relevant factors that make or break successful informational governance. But what is and how can we judge successfulness of informational governance as part of environmental reform? In the debates on e-governance and mandatory disclosure programs, three sets of criteria seem to circulate in defining successful informational regulation or governance: environmental effectiveness, democracy and economic efficiency. On none of the three criteria are easy answers possible. Although in general the TRI disclosure programs are credited with environmental effectiveness (see earlier), Grant and Jones (2004) are more sceptical regarding that of the right-to-know programs and laws of states in the United States, claiming that these had no significant net effects on plant toxic emissions. In their evaluation of twelve new voluntary, collaborative and information-based approaches in environmental policy – a slightly different category from what I have focused on in this book – Norberg-Bohm and de Bruijn (2005) conclude that the environmental effectiveness of these experiences is at best ambivalent. The debates around new digital divides and informational overflow in e-governance show similar ambivalences on the democratic dimensions of informational governance, whereas the direct costs – at least for regulatory agencies – may be lower. These ambivalences are expressed in calls for a better evaluation of the costs and benefits – not only in monetary terms – of various modules and forms of environmental disclosure arrangements (cf. Beierle, 2004). Others, however, remain convinced of their positive overall contribution, but stress the need for a better understanding of how these informational regulation programs work (e.g., Stephan, 2002). This discussion is not unlike the debate on environmental effectiveness of conventional environmental policy instruments and approaches during the past three decades, which have led to equally diverging conclusions and doubts.

The next two chapters will further elaborate on the continuities and discontinuities that come along with informational governance, with a stronger focus on private market parties and civil society, respectively.

7 | *Greening the networked economy*

1. Environment in a global economy

Ever since the emergence of the environmental crisis in industrialised countries, companies and producers have been blamed for both their significant contribution to environmental deterioration and for their limited efforts and responsibilities in decreasing and preventing environmental disturbances. But with the literature in the 1990s on ecological modernisation, on governance, on partnerships, on corporate social responsibility and the like, it has become commonplace to say that environmental protection is no longer the privileged domain of formal state policies and politics, and that increasingly private market actors become involved in environmental protection.[1] Although private producers and multinational companies are still major contributors to environmental problems, they can no longer be considered just passive environmental actors that move only when forced to do so by states and civil society. Companies and the private sector have developed their own agenda, approaches and organisational modes to engage – more or less successfully, more or less actively – in the environmental arena.

With the information revolution and globalisation the context in which and the way how companies operate has transformed considerably, as the wide literature on, among others, post-Fordism, the networked economy and the consumerist turn has illustrated. In this chapter, I especially will explore how these new conditions and structures of the networked economy have dramatically influenced the involvement and practices of market actors in environmental governance. How do

---

[1] There is still a debate regarding to what extent and how successfully private actors have emerged on the stage of environmental reform (e.g., the debates between neo-Marxist and political economy–inspired schools of thought, on the one hand, and various versions of ecological modernisation, on the other; cf. Mol and Sonnenfeld, 2000; Mol and Buttel, 2002).

market actors and companies produce and use information as a strategic resource in their environmental – or environmentally relevant – activities? How does this differ from conventional modes of environmental protection in the economic domain?

In exploring the role of information in moving towards – or obstructing – environmental reforms in the economy, three dimensions will be distinguished: (i) in-company use of information for environmental restructuring; (ii) 'private' informational governance of companies towards other market parties in economic networks and chains; (iii) informational activities and strategies of companies in relation to the wider political and civil society domains. Before elaborating on these three dimensions a short introduction will be given on the changing economic structure under processes of globalisation and the ICT revolution.

2. Informational economy

As with all claims of major social transformations, the idea of a significant structural change in the global economic structure and system during the past twenty-five years is meeting much debate and contestation. Not only are several leading economic, political science and sociological scholars questioning the idea of a major shift or rupture in the economic structure during the recent decades (cf. Thompson, 2004; Hirst and Thompson, 1996), those that agree on such a major transformation are debating the nature and characteristics of this change. Are the central organising principles of contemporary economic systems to be understood in term of post-Fordism, a networked economy, a global economy, an informational or knowledge economy, or any combinations of these ideas, such as, for instance, rephrased in terms of the New Economy? Theoretical arguments and empirical evidence of various natures are exchanged and mixed in these debates. This is not the place to review these discussions in all their details and refinements, let alone to draw final conclusions on the nature and extent to which the economy has been transformed lately. But the key tendencies of change can be summarised, which provide us at least with some insights in the direction of economic transformations and – in the end – the role information is playing in that. With Castells (1996/1997), most authors would agree on three major characteristics that are foundational of many of the economic transformations today: globalisation, networking and informationalisation.

Since the early 1990s, it has become mainstream to interpret and analyse the organisation of the economy in terms of commodity (or value) networks and chains, often transnational or global in character. After the Fordist mass production period, during which single companies were giving pride of place, in an era marked by globalisation networks and chains are the key units of analysis to understand the organisation of the (global) economy. Economic and industrial network analysis (Axelsson and Easton, 1988; Håkansson and Johanson, 1993; Grabher, 1993), global commodity chain analysis (cf. Gereffi and Korzeniewicz, 1994), the agro-food network and commodity complexes (Goodman and Watts, 1997; McMichael, 1996), convention theory (Ponte and Gibbon, 2005), and even the business systems approach[2] (Whitley, 1996; 1999) – while being in debate among each other – all focus on networks and chains to understand how the contemporary economy is (re)organised and works. At the centre of these studies and approaches lie the multiple linkages, relations and interdependencies of formally independent firms (or 'nodes'), which are organised increasingly on a global scale. These linkages are characterised by flows of material resources, finances, capital, knowledge and information. Of key importance in these chain and network studies are questions of (de)regulation and governance. In this literature, regulation usually refers to external state or intergovernmental rules that are imposed on – and thus structure – the networks and chains, whereas governance refers to questions of internal network and chain coordination and structuring.[3] Humphrey and Schmitz (2002) have redefined governance in

[2] For instance, Whitley (1996: 417) interprets the "world economy as generally weakly organized in terms of stable networks of information and material flows . . . and in terms of stable transnational institutions which could support and reproduce such networks."

[3] Usually a distinction is made between forms of coordination in chains and networks and modes of governance. Although coherent modes of governance refer to the entire chain, often formulated in terms of buyer- or supplier-driven chains, forms of coordination differ between segments in a chain or network (Ponte and Gibbon, 2005). Modes of governance were originally dichotomised in producer-driven chains and buyer-driven chains, putting the power emphasis upside or downside the value chain (cf. Gereffi and Korzeniewicz, 1994; Raikes et al., 2000). A general tendency is often identified in the global economy of power shifts towards the buyer-end of chains and networks, with captive suppliers in developing countries. The extent to which this actually happens depends on the commodity chain involved (cf. cars versus food), among others. The leading network-actor at the buyer side also may differ strongly among

global commodity chains and networks in terms of how certain (lead) firms set, measure and enforce the parameters under which other firms in the chain and network operate. Who is able – and how – to exercise control by specifying what products need to be delivered, in what quantity and when, how these should be produced and at what price (Ponte and Gibbon, 2005: 5)? Especially in the various contributions to the development of convention theory, quality conventions in commodity chains and networks have become central in the coordination between segments of the commodity chain.[4] And these quality concerns move beyond a narrow market logic of product quality characteristics, towards broader issues of process and production methods and – among others – the environmental concerns related to that.

In these (global) economic networks and chains, information and communication technology is increasingly a vital resource for economic agents and a structuring element. ICT is greatly facilitating information flows, which have become crucial in, for instance, detailing product specification, the organisation of production schedules and just-in-time delivery, and the monitoring of product quality and production standards. Through ICT and electronic infrastructure, business information of various kinds is swiftly transferred in economic networks, both from producers to suppliers and customers (B2B) and from producers to consumers (B2C). But ICT itself also enables and 'pushes' the restructuring of these economic networks and chains, in which parts of production steps and services can be relocated to the other side of the globe, without any serious consequences for product quality, costs and logistics.

Following Castells, Scholte (2000), McPhail (2006) and many others illustrate not only how the global economy becomes informationalised but also how communication and information capital and firms become globalised. ICT industries, telecommunication companies, (electronic) data processors and services, polling agencies, software

---

commodities, ranging from retailers, branded marketers and industrial processors to international traders.

[4] Conventions are defined as "shared templates for interpreting situations and planning courses of actions in mutually comprehensive ways that involve social accountability" (Biggart and Beamish, 2003: 444). As such conventions – whether they are purely economic or incorporate wider normative concerns – provide a basis for judging situations and actions, but at the same time form constraints.

producers and maintainers, mass media conglomerates and others have all taken a strong transnational outlook in their development. Paehlke (2003: 79) has been sharp on the role of media (in its various physical outlooks) in the current economy. "Most media are directed to a dominant objective – to increase, influence, and organise commercial transactions through advertising. Books and recording are perhaps the only communication forms that have thus far avoided being largely subsumed by this single-minded purpose." Advertisements are then seen as the central logic that structures the major share of the media, or even of contemporary culture.

Moreover, the market value of many physical products is far less than the 10 percent determined by the physical contents and a far larger share by its image content, a critique that is even more vehemently addressed by Naomi Klein's *No Logo*. In that sense, the informationalisation of the economy has even more far-reaching consequences, beyond the reorganisation of conventional economic networks and chains in which physical products are produced. In this line, one of the most radical analyses of an informationalised and mediatised economy is perhaps given by Scott Lash (2002) in his *Critique of Information*. If physical capital, accumulation and commodification were the key categories of the industrial economy, informationalisation, circulation and intellectual property are those of the new economy. Informationalisation is driving commodification, and "explodes the distinction between use value and exchange value," before being recaptured by capital for further commodification and profit-making. According to Lash (2005), Ford was commodified, Nike is mediatised.[5] The new consumer goods of the informational economy are characterised by their quick turnover (or immediacy), their largely immaterial nature, their global reach between production and consumption and their regulation through intellectual property rights. Actual production is outsourced, whereas the valorisation towards profit – design, branding and copy and property rights – remains in the core economies. This all results in new (and more nasty) global inequalities and power relations, according to Lash.

---

[5] Here, Lash refers to Baudrillard's (1981) notion of sign-value. The sign-value of an object refers to its capacity to generate communication, but can be and is converted into exchange value in an informational economy. Still, according to Baudrillard, the two are analytically different.

Being aware of the changes that ICT and the prominence of information bring about in the global economy, we nevertheless have to put these transformations into perspective: ICT is not (yet) turning the economy upside down as some of the radical New Economy adherents want us to believe. Nor does informationalisation make the economy virtual, weightless, dematerialised and footloose.[6] There is no 'sudden' change towards one dominant informational economy, but, rather, a mixing of old and new economic forms of production and consumption: informational configurations are parallelled by mass production, flexible specialisation and craft production (Thompson, 2004). And even in the informational configurations of the networked economy ICT cannot replace face-to-face interactions completely. Codified, routinised and scientific knowledge and information is easily transferred faceless, but tacit knowledge, trust, reliability and contracts need face-to-face, or handshake-based, interactions. Hence, this explains the ongoing logic of clustering and proximity of economic production networks (e.g., Porter and Ketels, 2003).[7] Thus, several authors conclude that distance still remains important in many economic networks and chains, and that, in some respects, the 'distance of trade' between countries is decreasing rather than – as one would expect in times of globalisation and ICT networking – increasing (Thompson, 2004: 567).

Against the background of such transformation and continuity in the global networked economy, the next sections analyse how information and information systems play a role in environmental protection in economic practices of production and – to a lesser extent – consumption.

3. In-company environmental management and public accountability

In his book *The Audit Society,* Michael Power (1997) explores the explosion of monitoring, verification and communication that has emerged in various contexts and systems in our modern society. During the 1980s and early 1990s, the word *audit* began to be used with increasing frequency, parallelled by the establishment of all kinds of new

---

[6] In *Globalization and Environmental Reform,* I have dealt more in detail with the materiality and groundedness of the global economy (Mol, 2001).

[7] Also within services and financial markets firms seem to concentrate and conglomerate in a few huge cities. See Sassen (1994) for a detailed analysis of the reason behind the geographical concentration of financial firms.

organisations and institutions: national audit offices, audit commis-
sions, audit guidelines and codes, private auditing companies, global
standards for auditing and so on. Moreover, the organisations and institutions started to shift their focus from conventional financial auditing
to all kind of other fields and domains, including the environmental domain.[8] And, indeed, in a substantial part of his book, Power
discusses practices, institutions and developments in environmental
auditing. The background of this auditing explosion is to be found in
new forms of governance and accountability at the crossroads of the
public-private divide,[9] and the growing amount, diversity in sources,
and complexity of information, without any clear undisputed authorities that validate (environmental) knowledge and information.

Environmental auditing – in a more narrow focus related to economic production – connects to all kinds of activities to monitor, assess
and certify the environmental management and performance of individual companies. Originally in the United States, environmental auditing
developed as an internal governance tool for the company management to monitor, control and manage the environmental quality and
performance of its various production plants, sites, lines and products.
The background for this development was to be found in both the
market, where credit institutions and insurance companies asked for
certainties on potential hidden liabilities (cf. Gunningham and Prest,
1993),[10] and the state, where increasing requirements from governmental agencies in licensing, monitoring and reporting and compliance
forced companies to standardise and organise their internal procedures
more systematically.

Especially in transnational corporations, with plant locations or
firms spread throughout the world, information systems are of key importance in harmonising company environmental governance and performance. Under various pressures, among which those of a globalising

[8] Power (1997) discusses the broadness and imprecise character of the category
auditing, which entails numerous activities that are often very different in
nature. But any stricter definition will exclude activities that are currently
labelled auditing, making such a more 'scientific' definition not very useful.
[9] See, for instance, the analysis by Stretesky and Gabriel (2005) of the 1995
Audit Policy of the U.S. EPA, where companies are encouraged towards
self-policy by discovering, disclosing and correcting their own violations of
environmental acts (in their case, the Clean Air Act).
[10] In an empirical study on the impact of two U.K. insurance companies on the
environmental performances of their clients, Minoli and Bell (2003) found out
that such insurance companies are still weak environmental regulators.

environmental movement and multilateral environmental regimes, transnational companies are increasingly challenged to formulate and implement one company environmental policy for all its locations, rather than differentiating environmental performance according to the local environmental regimes at the place of production. Uniform environmental reporting standards, companywide electronic information and communication systems, certified environmental auditing systems, common environmental training programs and companywide environmental auditing teams all contribute to that. Our investigations into the environmental policies and performances of four transnational companies, each with premises in Amsterdam, Beijing and Sao Paulo, revealed different degrees of, and policies towards, environmental harmonisation in these premises (Presas, 2005; Presas and Mol, 2006).

In these transnational companies, but also in national firms, ICT plays a major role in various steps in environmental auditing: (automatic) environmental monitoring; data collection, calculations and modelling; information handling and transmission; information reporting and exchange; and information verification. Many consultancy firms have specialised in developing software packages and informational systems for environmental management in industries and other private sector business. In most TNCs, continuous data collection, recording and access is becoming the standard, instead of the conventional idea of incidental monitoring and measurement of pollution. Modelling and estimations based on inputs and production schedules increasingly replace actual sampling and laboratory measurement of waste flows, not unlike that which we identified in state monitoring programs (Chapter 5). ICT and the Internet are also dominant in the production process of these environmental data and reports, the verification processes, the internal and external distribution, and the dialogues on environmental data, information and reports within and outside the company. As Isenmann and Lenz (2002) have argued, ICT and Internet use in corporate environmental reporting has four major benefits. First, the internal and external goals meant by reporting can be better addressed (easier reach of the target groups, easier to bring the corporate identity to the fore, better internal use and flow of the disclosed information). Second, environmental reporting becomes more efficient, as the internal reporting procedures and processes with respect to preparation, administration, distribution and presentation can be better handled. Third, the contents of the environmental reports are improved through ICT and Internet use, by – among

Second, auditing has been drawn into the regulatory framework, as part of broader changes in state-market relations and a reinvention of environmental governance. Although often the idea is that environmental auditing is related to regulatory relief, this is not clearly the case everywhere. In the United States, the Environmental Protection Agency has not been very willing to 'compensate' voluntary auditing with lower inspection levels or a 'managerial turn' in reinventing governance (cf. Rosenbaum, 2000). In some of the European countries, priorities in inspection do seem sometimes to be set in relation to certified environmental management or auditing systems, be it not always with positive outcomes.[12] Similar tendencies for regulatory control can be witnessed in, among others, industrialising countries in Southeast and East Asia and in developing countries in East Africa, where certified environmental auditing systems are more than incidentally criteria for lower priorities in inspection and enforcement programs. In addition, a regulatory space has been created for environmental auditing, where bottom-up company environmental management connects the internal working of an economic organisation with state regulatory programs. The initially voluntary company systems of auditing have increasingly found legal base in obligations regarding, for instance, environmental reporting. Denmark was the first European country with legislation on mandatory company environmental reporting in 1995 (with a revision in 2002), soon to be followed by the Netherlands, Norway, Sweden, Spain and others (Holgaard and Jørgensen, 2005). Of course, legislation on company environmental disclosures and reporting differs between countries, even between like countries (see, for instance, Nyquist's study [2003], comparing company disclosure legislation among Denmark,

whereas the final food product can only become HACCP certified if all production steps have received certification. In the 1990s, the Codex Alimentarius formulated general guidelines and incorporated HACCP into food hygiene codes. It is increasingly recognised by governments as a fulfilment of legal obligations towards food safety, within and outside OECD countries (Oosterveer, 2005).

[12] Several cases have been reported in which companies with certified environmental management and auditing systems met less stringent inspection regimes, but proved in the end to be among the most severe and illegal polluters. See, for instance, the mid-1990s case of Tank Cleaning Rotterdam in the Netherlands, which had a certified environmental management system but could nevertheless transport chemical waste from ships illegally to Belgium, to be dumped there as domestic waste (Stevens, 1999: 79ff).

Norway and Sweden), and calls for harmonisation (for instance, at the EU level) are regularly heard.

Third, environmental auditing systems have been strongly pushed by and become part of both business-to-business relations and civil society requests for transparency and information. As such these systems, and the internal informational processes and requirements coming along with that, have moved from a management tool for internal affairs to a requisite tool 'forced' on companies by their social and economic environment. Consequently, the informational flows related to and following environmental auditing are increasingly disclosed beyond the company gate. And although final decisions on environmental data and information transparency remain in most countries in the end part of company policies and interests, the external pressures make such decisions no longer a matter of 'internal affairs'. Several empirical studies (e.g. Patten, 1992; Deegan and Rankin, 1996; O'Donovan, 1999; 2002; Campbell, 2003) have compared company environmental disclosure policies within and across industrial sectors in OECD countries, using legitimacy theory. The general findings seem to show that companies that are more environmentally sensitive disclose over a period of time more environmental information than those in less environmentally sensitive areas, following the larger felt need to legitimise their company behaviour and safeguard their reputational capital. In addition, companies within sectors follow more or less similar disclosure policies, often also resulting from sectoral agreements.

Fourth and finally, these corporate environmental auditing, reporting and disclosure practices have moved from the OECD countries – mostly via transnational companies and the global networked economy – to domestic companies in industrialising societies. Jaffar et al. (2002) report, for instance, on the quality and quantity of environmental auditing, reporting and disclosure by Malaysian companies. Although still voluntary – and found in only some 20 percent of the Malaysian companies, with often ambivalent qualities and far from complete in terms of environmental indicators – these company environmental reporting and disclosures are clearly increasing in Malaysia (cf. Figure 7.1). Chapter 10 illustrates that the same is true for China but less so for Vietnam.

These advancements in company environmental information, and their reflection and codification in certification systems, also have other repercussions on companies in the lesser developed parts of the world. Although exporting companies in developing countries experience the need to upgrade their environmental management (and arguably performance) through these dynamics, there are also backsides and inequalities for third world companies. When ISO standards, with their informational requirements, are becoming the legitimised tickets (or barriers) for global market access, this especially affects domestic developing country producers. As Di Chang-Xing (1999) has analysed for China, governments and local industries often do not have sufficient experience and capacity to implement (certified) corporate environmental information and management systems and the costs of introducing such systems, as well as the independent auditors to certify them, are relatively high. Developing country governments and companies (and environmental NGOs) have been – and still are – only marginally involved in the drafting of, for example, the ISO 14000 series – and come often too late to make an effective contribution (Krut and Gleckman, 1998; Clapp, 2005).[13] All of this makes it at least understandable when some developing country representatives interpret informational requirements, such as included in ISO standards, in terms of protectionism and trade barriers, rather than as a fruitful basis for global environmental reform. And it does clarify why some scholars criticise these systems as 'privatization of environmental governance' (cf. Clapp, 2005).

4. Private governance in economic networks

Lately, especially convention theory (cf. Ponte and Gibbon, 2005) has reconceptualised the governance perspective on global value or

---

[13] For instance, of the 141 developing countries, only 50 were full members of International Standard Organisation (ISO) in the late 1990s and only 25 participated in Technical Committee 207 on environmental management (and its various subcommittees), which deals with most environmental standards (only five to six played an active role in the Committee until 1995 and a few more in 1996 and 1997). In contrast, all developed countries were members and almost all were actively involved in TC 207 negotiations from its establishment in 1993 onward (cf. Krut and Gleckman, 1998: 40–62). In 2006, the situation improved: 103 countries were members of ISO, and 69 of TC 207 of ISO (of which some 35 are developing and transitional economies). For similar data on the Codex Alimentarius standards, see Henson and Jaffee (2007).

commodity chains in two interrelated ways: (i) from issues of direct control and power, to more indirect modes of governance and coordination (control at a distance); and (ii) from a purely or dominantly economic, productionist perspective of commodity chain governance to more explicit attention to (product and process) quality. It is especially in such approaches that environmental governance finds an easy place, and that the role of information in private environmental governance takes shape.

Product safety arrangements and green conventions[14] can take different forms in governing product chains and networks. Some scholars still claim the proximity and the length of a product chain/network to be essential for quality and environmental conventions (e.g., those working in the tradition of local food networks). But this is only one mode of private quality governance. The identity (and thus quality) of a product can be guaranteed or institutionalised in the 'repetition of history' via three modes: through its region or country of origin, by its brand name or simply through repeated interactions between actors, irrespective of the length of a chain/network or its localised origin. In addition, knowledge of quality becomes increasingly embedded in standardised arrangements, often expressed through symbolic tokens such as (technical) standards, labels and codes of conducts, which reduce the need for other forms of quality institutionalisation. So, in general, as Murdoch and Miele (1999) argue for the food sector, there is no automatic link between quality and green conventions, on the one side, and specific (that is, localised) forms of intra- and inter-firm quality coordination in chains and networks, on the other. In the emergence of these different quality conventions, market signals and logics increasingly proved to be insufficient and are complemented by wider information exchanges. With respect to these forms of standardisation of quality conventions, and the informational flows related to it, lead firms have unequal power to control them. The political economy of networks and chains shows us that they have a more than equal influence and control on designing these standards and governing these flows, and thus on 'regulating' entry barriers to these

---

[14] In convention theory, several classifications are made between types of conventions, such as between market, industrial, civic and domestic conventions (cf. Boltanski and Thévenot, 1991; Ponte and Gibbon, 2005: 19).

conventions. But lead firms are by no means in full control of green conventions.[15] These conventions relate to much broader and fundamental sociostructural developments and domains, shaped at least partly independently of the wishes – and beyond the control – of lead firms.

It is especially for reasons of quality, safety and environment that commodity chains and networks have been pressed – by actors and institutions inside and outside these chains and networks – to increase traceability, transparency and trust.[16] This development relates strongly to what could be labelled private informational governance. In addition to the conventional flow of materials, products and goods downward in the chain, and the reverse flow of money upwards, the – third – flow of information has particularly increased following questions and public debates on quality, safety and environmental consequences of products and production methods. Especially in the agro-food sector with its natural basis, often open primary production systems, and transnational food safety crises, but also in other commodity chains such as that of health care (cf. Nouwt, 2004), information requirements have skyrocketed. This has resulted in the development of advanced systems of tracking and tracing,[17] strongly enhanced by governmental policies and legislation, such as the European Union's

---

[15] This is further elaborated in the 'economy of quality' (cf. Callon et al., 2002).

[16] Often a distinction is made between public market transparency and private market transparency. Public market transparency relates to consumers, whereas private market transparency is related to organisations in the commodity chain/network and is believed to have adverse effects on competition (Nouwt et al., 2004).

[17] Tracking refers to following products flowing through the chain/network. Tracing refers to finding out ex post how a product moved through the chain/network, from its constituent parts to the final end product. The European General Food Law uses a far-reaching definition of traceability (Art. 3): "the ability to trace and follow a food, feed, food-producing animal or substance intended to be, or expected to be incorporated into an food or feed, through all stages of production, processing and distribution." A comparative study across seven countries (Australia, Germany, the Netherlands, Spain, Sweden, the United Kingdom and the United States) in four supply sectors (meat, dairy, fruit and vegetables, and grain/bread) illustrated the poor spreading of traceability systems in the early 2000s (Vorst et al., 2003). This corresponded with investigations by the Food and Veterinary Office of the DG Health and Consumer Protection of the European Commission.

General Food Law.[18] In addition to governmental pushes, private company initiatives such as product stewardship policies, accountability programs, insurance requirements and social and environmental corporate responsibility programs have significantly contributed to ideas of tracing and tracking of products. Where information on (different steps in) the product life cycle go beyond the internal affairs of private actors in the chains and networks and are made public, tracing and tracking flows over into transparency. Information becomes then relevant for stakeholders outside the production systems: consumers, citizens, public authorities, NGOs and so on. Transparency aims to provide access to product-relevant information (in its broadest meaning) by those not directly involved in the commodity chains or networks. Although transparency was arguably initiated to satisfy worries of public authorities, private consumers and civil society actors, it also has important functions for the commodity chain itself, for instance, in preserving product identity, adding value and identifying (or preventing) illegal competitors.

Especially when transparency is related to actors outside the commodity chain, trust is crucial. In a globalised market, with increasing geographical distances in chains, trust has to be built increasingly via abstract information systems and symbolic tokens and less via sensory experiences, face-to-face communications and past records (or the "repetition of history"). Questions of reliability and verification of informational symbolic tokens become central. But also in large-distance linkages and coordination *within* commodity chains and networks transparency and trust via abstract information systems and symbolic tokens play important roles. Although transparency has become to a certain extent fashionable in entering the new millennium and practices of information disclosure in commodity chains and networks are increasing, there still exist many good reasons for, and continuing practices of, keeping (environmentally relevant) information

---

[18] EC regulation 178/2002 of the European Parliament and of the Council of 28 January laying down the general principles and requirements of food law, establishing the European Food Safety Authority and laying down procedures in matters of food safety. Consumer and environmental NGOs are even asking for further legislation in the area of transparency. For instance, the Dutch consumer organisation Consumentenbond stresses the need of and lobbies for an International Consumer Right-to-Know Act, comparable to the Freedom of Information Act in many countries (Peters, 2004; for a proposed draft of such an act, see http://www.consumentenbond.nl, accessed April 2006).

internal, secret and out of reach of private suppliers and customers/ consumers, public authorities and civil society. And there are also various legal barriers that restrict free-floating and flowing of information within commodity networks and chains.[19]

With the emergence of ICT, information systems in commodity chains and networks have been partly digitalised, both when information is applied for internal governance of commodity chains and networks (or private market transparency) and when information is related to external or public market transparency and trust. With that digitalisation, the possibilities of (private) informational governance are expanding. ICT greatly facilitates chain coordination effort and the exchange of information upstream and downstream. But it is also directly linked to monitoring and tracking products through their commodity chain at real time. To a significant extent, this digitalisation of chain information systems is related to logistics, product qualifications, just-in-time delivery, better chain coordination and so on, and less so to environmental and safety issues (see, for instance, the strong ICT-based information systems being developed in the health care chain). But, as argued earlier, following convention theory, increasingly environmental and product safety requirements are being exchanged in commodity chains, and ICT plays a growing role in that, too. There is an expansion of Web sites with product information and product comparison on environmental and product safety dimensions, an experimenting with webcams revealing production circumstances,[20] and a growing digitalisation of tracing and tracking systems.

Also in their encounters with critical citizen-consumers and environmental and consumer organisations, producers increasingly turn to the Internet. When NGOs have built informative and widely visited Web sites on companies with poor environmental, labour and/or social performances, companies are copying this strategy. Interactions with consumers and debates with critical NGOs take increasingly place via

---

[19] Some of these restrictions include intellectual property rights, privacy rights, liability concerns around making information available and competition laws regulating information exchange agreements (see Nouwt et al., 2004).

[20] See, for instance, Peter's Farm (http://www.petersfarm.com), an initiative taken by the Dutch calf feed provider Alpuro. Consumers of calf meat obtain a code that enables them to enter a Web site with webcam images of stables from which their calf meat originates. Similar initiatives with revealing Web sites, although not often with webcams, are found for other products.

weblogs of – or sponsored by – these companies. McDonald's, a company that has been on the frontline of public criticism on animal welfare, environment and healthy food, recently (2006) started a weblog to enter digitally into dialogue with its consumers and critics.[21] But other companies, such as Walmart, General Motors and the Japanese car company Mazda,[22] have a less open weblog strategy. They secretly mobilise (and reward) unidentifiable individual bloggers to promote their products, counteract (environmental) criticism and 'serve' other bloggers and media with positive company news, also with respect to environmental performance.

Agro-food and ecolabelling

As an illustration of private and public informational governance in the networked economy, I will detail the agro-food system and the exponential use of labels in greening agro-food networks. Conventional research in the Western agro-food sector has focused on the decline of the Fordist food regime of mass food production/consumption and the emergence of 'quality' and environmental issues in restructuring food production-consumption chains. There is probably no better sector to illustrate the role of private informational governance, as food networks are – and have to be – most advanced in issues of transparency and trust. Various (inter)national food crises, controversies and scandals (e.g., BSE, mouth and claw disease, bird flue, GMOs, dioxins in chicken meat, pesticide residues, hormone residues in meat, animal welfare) in national and transnational agro-food sectors and chains have, among others, resulted in all kinds of certification, standardisation, information, labelling and control systems, as well as in a growing market for green, organic or ecological food (experiencing similar informational systems). Some of these systems come from the sector (e.g., EUREP/GAP, BRC, GMP, SQF, ISO standards),[23] some

---

[21] See http://csr.blogs.mcdonalds.com. Other examples are http://blogs.ford.com of the car company Ford and http://blogs.msdn.com of Microsoft. There is a growing literature on company blogs, for example, Scoble and Israel (2006) and Holtz and Demopoulos (2006).

[22] Cf. *NRC Handelsblad*, 20 April 2006, p. 18.

[23] Euro Retailer Produce Good Agricultural Practices (EUREP/GAP), Good Manufacturing Practices (GMP), British Retail Standard (BRS) and Safe Quality Food (SQF).

are based on governmental laws and policies (in Europe, especially the General Food Law, obligations for traceability and product labels in among others the meat and GMO sectors,[24] European environmental labels), some are pushed and designed by civil society NGOs, and some find their origin in combinations. Consequently, there is an increase in tracking and tracing systems, and information and labelling in food chains and networks, strongly supported by tailor-made (but not yet very standardised) ICT systems.[25]

Organic labels and standards coordinate and govern one of the most dynamic and fast-growing agricultural markets[26] in the industrialised world, and increasingly also beyond that (see Chapter 10). With annual growth percentages often between 10 and 20 percent in the United States and in the EU, organic food markets are increasingly seen as an interesting niche market (Raynolds, 2004: 732). The International Federation of Organic Agriculture Movements (IFOAM), established in 1972, has now seen its definition of organic being accepted within more than one hundred countries. Its most important governance strategy is the international promotion of certification systems to regulate organic quality and consolidate markets. Although initially promulgated by IFOAM and national private voluntary certification organisations, these IFOAM organic standards and certifications are increasingly included in governmental regulations and guidelines (e.g., the EC regulation 2092/91 on organic production methods; the guidelines of the Codex Alimentarius).

It is difficult to get full overviews on the actual amount of environmental, green or organic labels prevailing on the agro-food or any other markets. An inventory of environmental labels on the Dutch consumer market in 2001 (Waart and Spruyt, 2001), found around one hundred

---

[24] On traceability and labelling of GMO, see EC regulation 1830/2003, *OJ* L 268, 18/10/2003, p. 24. For an introduction to the EU General Food Law (EC Regulation 178/2002), see van der Meulen and van der Velde (2004).

[25] To support the further foundation of ICT in food tracing and tracking, the Netherlands Ministry of Agriculture, Nature Protection and Food Quality, and a number of larger Dutch food industries set up the Platform Transparency and ICT in 2003.

[26] The world market for certified organic foods was estimated to be worth US$23–25 billion in 2003 and grows roughly 19 percent annually (Kortbeck-Olesen, 2003). Between 1985 and 2001, the average annual growth rate of acreage being brought under certified organic food production in Europe was 27 percent (http://www.organic.aber.ac.uk/statistics/euroarea.htm).

Table 7.1. *Number of environmental labels for different products or categories on the Dutch market in 2006*

Product/category	Number of labels
Food general	35
Meat	14
Fish	3
Eggs	7
Textile	10
Packaging	13
Recycling	29
Travelling	2
Electricity	9
Gardening	2

Note: Duplications not excluded.
Source: http://www.keurmerken.info, accessed September 2006.

official labels that refer partly or completely to environmental qualities of products or production processes. Besides that, more than 110 visualisations or logos were identified that referred to green or environmental qualities, as well as some fifty trade names with 'eco', 'bio' or 'nature'. Table 7.1 provides data on the amount of green or ecolabels for different product categories in the Netherlands in 2006, illustrating the diversity in numbers for different product categories.[27] Figure 7.2 provides an indication of the rapid growth in environmental labelling in the 1990s, by showing the increase of (especially local, national, regional) environmental labels, certificates and standards on the European tourist market from 1990 to 2000. Although a significant increase took place in that decade, in the first years of the new millennium some consolidation and integration of various labelling systems took place, resulting in some fifty-four ecolabels in 2005.

The differences in these ecolabelling and product-information programs deserve some further attention. Most advanced industrialised states have now one or more state-recognised and sanctioned

---

[27] See http://www.eco-labels.org, a Web site of the U.S. Consumer Union, for an overview of environmental labels on the U.S. agro-food market. This U.S. Web site mentions forty environmental labels for meat, two for fish and thirty-one for eggs (accessed September 2006).

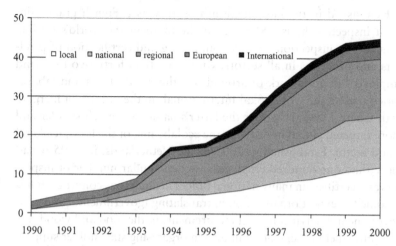

Figure 7.2. Number of environmental certificates and standards for the tourist companies and accommodations used on the European market, 1990–2000 (*source*: http://www.eco-tip.org/Eco-labels/ecolabels.htm, accessed September 2006).

ecolabelling programs for various products (e.g., on organic food, wood, recycled products, cars, green energy, etc.), and incidentally international state-run ecolabelling programs are being developed (e.g., in the European Union). But in addition to these state-sanctioned and often state-organised ecolabels and product information schemes, a blossoming of private initiatives can be witnessed from local or national producer/sector organised initiatives, to truly global innovative industry-NGO ecolabelling initiatives such as those of the well-known Forest Stewardship Council (FSC) and Marine Stewardship Council (MSC).[28] The environmental labelling and certification schemes have become a worldwide phenomenon, in which global networks of certifiers and inspectors should make sure that information flows represent what they claim to represent.

---

[28] A rich literature from various disciplines has emerged on the various forms, impacts, and arrangements involving public, private and mixed labelling schemes (cf. Magat and Viscusi, 1992; Oosterveer, 2005). But the global labelling schemes are not always as global as they are meant to be. The innovative and much celebrated MSC and FSC labelling schemes are developed to function globally, but in 2006 seventeen of the nineteen certified fisheries under the MSC label and more than 80 percent of the certified area under the FSC label were located in developed countries.

Established in the late nineteenth century as a French grain ship-
ment inspection house, SGS[29] is at the moment the world's leading
company in inspection, verification, testing and certification of products and processes in all sectors, with more than forty-two thousand
employees in a network of around one thousand offices and laboratories around the world. Skal International (in the process of merging
with Control Union[30] under the latter's name) is a much smaller and
more targeted global network in the ecolabelling of the food, feed and
wood sector. Established in 2002 in the Netherlands, in 2005 it had
more than one hundred employees, with a similar number of inspectors and certifiers in many national offices in numerous countries. It has
become the expert organisation in translating (governmental and nongovernmental) environmental requirements in the food and wood sectors into labelling schemes, but also in organising stakeholder support
and influence, in assessing what can be and what cannot be certified
and inspected, in product inspections (along the whole line of the value
chains and the transport routes). They run various programmes such
as EUREP/GAP, Greengold label (on palm oil use in biomass energy
production), Tesco Nature's Choice, Utz Kapeh coffee[31] and organic
production.

Each label and product-information system – whether it is in the food
or any other sector – has its own arrangement of actors collecting the
information, organisations verifying information, routes for transmission of the label or product information, arrangements and strategies
to deal with uncertainty and trust, and users that are addressed with
information. This diversity and omnipresence of labels and product
information, as well as the diversity of arrangements that come along
with them, have triggered fierce debates and controversies on verification, public and private responsibilities, scientific basis, effectiveness
and trade barriers, without, however, jeopardising the further development, growth and importance of labelling practices and related information flows until now. The strong controversies also give evidence of

---

[29] See http://www.sgs.com/.

[30] See http://www.controlunion.com/main/default.htm.

[31] Although inspection and verification usually goes down the value chain,
making sure that information of upstream activities (e.g., on production
circumstances) comes down to the end producer or retailer. In the Utz Kapeh
coffee scheme, the innovative inspection goes upwards, making sure that the
one-cent additional price paid by the consumer is returned to the individual
small-scale coffee producer.

the more than marginal role these informational governance arrangements around labelling play, according to the various stakeholders (cf. Overdevest, 2005).

Similar to the production process standardisation of the International Standard Organisation (ISO), these environmental product labels meet usually reluctant developing countries on their way. The conditions for applying such labels have often been set by the largest players without much participation of developing countries, and usually it is especially the larger, well-resourced producer networks that are able to institutionalise the arrangements necessary for fulfilling these labelling requirements. In a detailed empirical analysis of various ecolabels, Ponte (2006) shows how, among others, the MSC and the FSC labels, for fish and forests, respectively, have generally offered the more sophisticated and larger suppliers an advantage. Most of the labelled fish and forests relate to the large companies and the developed countries, and not so much to the small-scale domestic ones in developing countries. These labels are expensive to comply with, but in time will become the minimum standards for the market. In general, these sustainability labels have moved to "hands-off, auditable, systemic and managerial approaches to sustainability" (Ponte, 2006; 48), in which expert knowledge, scientists and system managers are key actors that look for conformity to system performance and specific rules, rather than to stated objectives of sustainability, safe food and fair trade. Thus, Ponte (2006) argues, special systems of compliance and verification are needed to cater to the needs of the developing countries and small-scale producers, instead of just financial and technical assistance for them. Labels and labelling processes are unequal power games, which are not repaired by simple assistance.

The strong increase of labels, and the growing debates and controversies about them, illustrate the importance of informational politics and the role that labels play as symbol mobilisers and trust generators. What happens with labelling replicates – but at the same time may contest – the branding approaches of private companies, in which billions have been invested. With the growing importance of branding, sign-value and a mediatised economy (Lash, 2002), one can expect that ecolabelling and environmental symbolism increases. Although environmental NGOs generally have supported critical analyses on branding by MNCs (with Naomi Klein's *No Logo* as the leading view), they also have learned the importance and power of symbols, trust and their own logos. The 'marketing' of WWF's Panda logo is arguably the best example of a new role of environmental NGOs in the informational politics of the networked economy. It gives evidence that legitimation and reputation, codified in a symbolic and informational 'instrument' as a logo, has power. But similar to brands, the power and value of such NGO logos also can be damaged, and with that the reputation and legitimacy of the NGO, as will be further explored in Chapter 8.

5. Monopolies, distortion and public relations

With the increasing importance of information in environmental governance within and between private producers and consumers, we also should be aware of the weaknesses, strategic company behaviour and problems that come along with it. What are the dangers and downsides of private or company informational governance? With the emerging importance of information in regulating environmental performance questions of access and transparency, verification and control, uncertainties, presentation, liability/accountability and information misleading become crucial. Some of these downside aspects run parallel to what we have identified in the previous chapter of state-related informational governance; some problems are specific for the private sector.

First, as has been noted earlier, in informational governance access to information is crucial. Notwithstanding the developments in company environmental reporting, tracking and tracing and transparency that can be witnessed in various sectors and commodity chains, and the increasing codification of that in laws, access to relevant environmental information for private environmental governance of economic actors, chains, and networks is far from routine practice. It is more than incidental that information is not available, restricted in access for various reasons, or impossible to relate to specific companies. A too-heavy reliance on informational governance with limited access and transparency rebounds negatively on environmental reforms. This is even more so when information is spread unequally, both within economic networks and chains and between these networks and chains, on the one hand, and public agencies, consumers and nongovernmental organisations, on the other. In that sense, (new) regulatory arrangements on information access, transparency and verification are essential for supporting well-functioning private informational governance.

But equally important are nonregulatory developments that push increased transparency and access of environmentally relevant information for all stakeholders (e.g., the importance of trust, reputational capital and corporate social responsibility). The right-to-know and disclosure legislation, which were and are crucial in opening the public domain on information, should be parallelled by similar – legal or nonlegal – initiatives in the private domain.

Second, and related, informational governance within the economic domain is restricted as a result of the limitations on transparency that are set by legal regimes, competition law and privacy rules. Within capitalist market economies and democracies that cherish individual rights of privacy, policy and legal systems usually limit the collection and dissemination of environmentally relevant information. Information sharing between competitors may run against policies of competition authorities. Equally, information collection on consumers and customers might be obstructed by privacy regulation. Information collection and use of citizen-consumers on energy use, mobility behaviour, shopping and water consumption may be essential for informational governance in commodity chains and networks, but often are believed to run against the privacy regulations of individuals. This puts – again (see Chapter 5) – the question of surveillance on the table.

Third, and closely related to questions of access, are problems on reliability of information. A major share of information is still produced by, in the hands of, or controlled by a few dominant players in the economic networks and chains. Those players do not always have an incentive towards complete and reliable information (disclosure). Auditing, verification and control, and certification of information become then essential. But audits, verifications and certifications are complicated, time-consuming and often depend on cooperation with those in charge of information. And since the 2002 scandal of Enron and Arthur Andersen,[32] we also have to worry about who is controlling the controllers and how reliable is that. The possibilities of generating incomplete, incorrect, irrelevant or misleading information are far from negligible and can thus seriously endanger private informational governance, for instance, around the proliferation of green

---

[32] Since the start of the new millennium, KPMG, Deloitte & Touche, Ernst & Young and PriceWaterhouseCoopers also have been involved in various accounting scandals, similar to the Enron/Andersen case.

labelling and certifications. Although in most countries there exists one official, often state-recognised, ecolabel, there is at the same time a diversity of labels, visualisations, product-information suggestions, logos, symbols and colouring of product packaging that aims to communicate a message of environmental friendliness. Sometimes, these are backed by private interest governance (Jacek, 1991) standards, sometimes these are controlled by independent institutions, and sometimes these are individual initiatives of one or a group of companies. The reliability issues that come along with some of these may undermine trust in the entire labelling and certification systems, and thus limit the informational power in these symbolic tokens.

This all is further complicated with the significant uncertainties that come along with environmental knowledge and information (cf. Chapter 3). Life cycle analysis, modelling, risk assessments and other natural science tools all aim to rationalise, objectify and aggregate environmental information related to products and production. But the natural sciences and their tools have not been able to fundamentally take away problems of perceived uncertainties with respect to, for instance, climate change, GMOs, mad cow disease, or toxic substances in plastics and food. These uncertainties work two ways in private informational governance. In some cases, uncertainties may hamper and disable public governance by law, increasing the (temporary) emphasis on (especially private) informational governance (such as, for instance, in the European case of GMO labelling). Equally often, the absence of 100 percent certainty is used strategically to further distort and frustrate environmental governance through what we might label informational nongovernance, for instance, in green backlash campaigns in the United States (cf. McCright and Dunlap, 2000; 2003). In both cases, conflicts of interest and ideas are increasingly fought on the battleground of environmental information, especially because the inherent uncertainties seem to allow fierce and enduring battles without any authoritative judge.

Fifth, and finally, environmental information and communication in economic systems are also characterised – or, sometimes, dominated – by advertisements, marketing and public relations. With the growing – also economic – importance of environmental and health issues, advertisements, marketing and public relations increasingly include environmental references. With that environment becomes a marketing tool, strategically used in large-scale and forceful advertisement campaigns and company symbols. Systems of tracking and tracing, corporate environmental reporting and life cycle analyses then become blurred with multimedia advertisements, whereas entirely different rationalities, logics and forms dominate. Although this boosts the informational economy, it might seriously jeopardise private informational governance of the environment. The question then becomes if systems are in place to discriminate these strategic advertisements from truly informational governance on the environment. Notwithstanding the developments in internal codes of conducts, social corporate responsibility and the like, such 'revealing' and discriminatory systems have often to be found outside the economic networks and chains.

6. Conclusion: stateless governance through information?

In the global networked economy, environmental and sustainability issues can no longer be ignored by private firms. They have become part and parcel of their business environment, whether it is in the form of new (private and public) regulations with which they have to cope, new market opportunities and competition areas or new challenges put forward to them by (organised) citizen-consumers. In general, private actors have understood these new conditions quite well, and new governance theories have reflected on the growing involvement of private actors in environmental governance. This chapter has analysed to what extent and how information has become increasingly important in environmental business strategies and activities. Within companies, between companies, between companies and consumers and in state-market interactions information has become a key resource in dealing with environmental challenges and pressures. And information politics has become a key domain in which environmental controversies are at play. Although we should acknowledge and be aware of the growing importance of information politics in environmental controversies that involve private producers, we should not be too naive regarding the final contribution of such private informational strategies and activities to sustainability.

In reviewing the emerging tendency of stateless regulation of transnational economic networks through the monitoring of transnational companies by consumers and companies themselves, Seidman (2005) is quite straightforward. The monitoring and informational activities of civil society and consumer NGOs, on the one hand, and corporate responsibility and codes of conduct programs, on the other, are by far not sufficient to correct the negative social and environmental side-effects of a globalised economy. State involvement is essential to turn nongovernmental monitoring and informational activities into on-the-ground improvements. Stateless governance through information is not a model that will work on its own, which puts our analysis of private informational governance again in perspective. Kerwer (2005) adds to this by arguing that the widespread emergence of private, voluntary, global standards raises the issue of accountability and relating such voluntary standards to formal state directives is one way to deal with accountability in a still largely state-organised global governance system.

At the same time, as Seidman (2005) illustrates with the cases of the Sullivan Principles, the Rugmark labelling program and the Commission for the Verification of Codes of Conduct, these informational governance arrangements beyond the state are new forces in a global networked economy, addressing transnational practices beyond the nation-state border and as such playing a role in 'taming globalisation' (Held and Koenig-Archibugi, 2003) in a new era. Although the 'transformative powers' of stateless governance along these private informational lines are indeed limited, this chapter has argued that we can – and should – no longer ignore the nonconventional regulatory dynamics that are emerging through private informational governance arrangements in the networked economy. Private informational governance can contribute to environmental reform because it is part of and embedded in wider systems of state regulation, transparency, legitimacy and reputation. Especially in the latter three, the environmental movement plays a major role.

8 | Environmental activism and advocacy

1. A natural alliance in transition

On 19 August 2006, a ship called the Probo Koala – chartered by the International petroleum trader Trafigura LTD and controlled by the Dutch firm Trafigura Beheer BV – unloaded more than five hundred tonnes of toxic waste[1] in Abidjan, the main economic capital of Ivory Coast in western Africa. Before that, it tried already to unload its waste in the port of Amsterdam, but waste treatment proved to be too expensive (around US$250,000) and too slow (leading to another US$250,000 loss following late arrivals). The ship moved to the western coast of Africa where it could unload its toxic waste at a waste processing company calle Societé Tommy, which was established just a few days earlier for only this purpose[2] at a price of around US$15,000. The toxic waste was subsequently dumped at more than ten sites in the African city, leading to medical treatment of thousands of victims with symptoms including respiratory problems, nausea, dizziness, vomiting (including throwing up blood), burns and irritation. Following local media reporting on these symptoms, local unrest started. It was not until the first week of September 2006 that the incident became globally known through the networks of activists and media. The Ministry of Health of Ivory Coast announced an extraordinary meeting that led to the dismissal of its government in September 2006, to be replaced by almost the same cabinet a few days later (only the ministers

---

[1] The disposed waste consisted of liquid sludge containing large quantities of hydrocarbons, contaminated with at least three substances: hydrogen sulphides, mercaptans and caustic soda (as reported by the Centre for Anti-Pollution Control in Ivory Coast).

[2] Investigations focus not only on the criminal role of Societé Tommy, which unloaded the toxic waste and dumped it, but also on the company Puma Energy – a Trafigura subsidiary – that stocks oil products in Abidjan and granted the contract to Societé Tommy, and Waibs, a company that handles formalities for ships docking at the port.

for Transport and Environment were replaced). International reactions followed swiftly, including international scientific inquiries, legal cases against Trafigura and medical and environmental aid to the city of Abidjan.   •

This example illustrates various well-known stories with respect to failing authorities (in Amsterdam and Abidjan), the global political economy of waste flows and waste dumping, the devastating consequences of unregulated global capital and TNCs, the poor implementation of multilateral environmental agreements, and the global networks and flows of activists and media that can in no time activate states, international organizations, and public opinion, with significant consequences for the powers that be. It is the environmental NGOs that are of interest here. And with respect to the flows and networks of activists and information, one is tempted to conclude how little seems to have changed over the past two decades.

Since the early days of their emergence in the 1960s and their institutionalisation into NGOs in the 1970s, environmentalists have used information as one of the key resources in their struggles with states and economic actors. Although initially scientific information was of key importance in getting, among others, pesticides, nuclear energy, and water and air pollution on the local and national public and policy agendas, later their information politics spread much wider: beyond collecting and disseminating scientific information, beyond a focus on the local and national levels and beyond state actors and politics. Throughout the 1990s, one also can witness a further professionalisation in handling informational resources and strategies by environmental activists and NGOs. Communication and information specialists and strategies increasingly became an inherent part of NGO campaigns, or even moved to the very centre of campaigns. Whereas Greenpeace, with its mediagenic and visually spectacular campaigns, had understood the role of information and communication already quite early (e.g., Hansen, 1993a; Wapner, 1996: 41–71), today most environmental NGOs are fully aware of the essence of what we would label 'informational governance'. To be sure, compared to the other actors in contemporary informational politics and governance, environmentalists were much earlier and for a long time more advanced and concentrated on informational resources; strategic open and off-the-record disclosures; shaping mainstream media agendas through briefings, leakages and high-profile press conference; and mediagenic and visually attractive and spectacular campaigns. Their early focus on information and informational resources also was caused by their limited access to and control over economic and authoritative resources. Gamson (1995) even claims that environmentalists are media junkies, being so strongly oriented to, and measuring the effects of their (won) campaigns by, media coverage. In all, environmental NGOs have been better and especially much earlier aware of the importance of reputation, legitimacy, media coverage, trust and the use of symbols than the later actors in informational governance: state agencies, the private sector and the 'old' social movement actors. The 1995 battle between Greenpeace and Shell on the planned – but finally prevented – dumping of the oil platform Brent Spar in the Atlantic Ocean was not a battle fought in the Atlantic but a media and information battle. It illustrates the advanced informational politics of Greenpeace compared to a multinational corporation – and the U.K. government – that relied and focused too much on conventional power struggles. In that sense, there is an almost natural alliance between environmental activists and NGOs, on the one side, and informational governance, on the other.

There is no need to repeat the analyses on the ways in which environmental NGOs innovatively dealt with and applied information and informational resources during the heydays of conventional environmental governance (see Box 8.1 for an example). Others in the tradition of new social movement research have done so in much detail. In this chapter, I will focus especially on how and to what extent the natural alliance between the environmental movement and informational governance is changing under conditions of globalisation and ICT. With the growing importance of information and communication technology, significant changes in the information politics and practices of environmentalists and the global reach of environmentalism and the more advanced informational politics of states and the private sector can be witnessed on three fronts: (i) internal organisation, communication and structure; (ii) strategies, campaigning and coalition and (iii) the enhanced geographical scale and outreach. These three domains of innovations in civil society's informational governance will emerge from our analysis in this chapter, and we will conclude with an analysis of the dangers and vulnerabilities of high-profile informational politics for environmental activism.

8.1 Informational governance in the electronics industry

The global electronics industry is a notorious polluter. Several examples highlight the emergence of informational governance by environmental NGOs in cleaning up this sector (Smith, Sonnenfeld and Pellow, 2006).

The Silicon Valley Toxics Coalition has been one of the earliest NGOs challenging the electronics industry to clean up. In various campaigns, it was an early adopter of cutting-edge electronic tools, such as the development of Internet-based interactive ecomaps documenting the pollution levels caused by the electronics industry in Silicon Valley and the export of pollution globally.

In Taiwan, pollution of the electronics industry has been challenged by community-based monitoring systems. The community-based Environmental Supervision Network trains local volunteers to monitor high-tech pollution with the help of university scholars. Villagers and farmers are trained to monitor pollution around Hsinchu Science Park, following the lack of publicly available data and information on emission and pollution levels (Chang et al., 2006).

In the United States, the Computer Take Back Campaign (CTBC) developed in 2001 a system of 'Computer Report Cards', which scores electronic companies on a large number of environmental criteria. Since then, these scorecards have been renewed regularly, and the computer industry has been quite responsive to these informational campaigns. Several of the companies felt enough pressure to take action and enter into negotiations with the CTBC. Wood and Schneider (2006) report how this affected Dell, one of the market leaders in personal computers.

More recently, in 2006, Greenpeace developed a Green Electronics Guide, ranking the various electronics companies (especially with respect to personal computers and cell phones) according to the environmental friendliness of their products. This soon resulted in companies such as Hewlett Packard and Dell making a commitment to improve the environmental performance of their products with respect to brominated flame retardants and PVC (http://www.greenpeace.org/international/campaigns/toxics/electronics/how-the-companies-line-up).

2. Digitalising environmental NGOs

The digital revolution and the spreading of information and communication technologies in the late 1980s changed environmental activism in the developed part of the world. The various investigations on the use of Internet and ICT in green activism illustrate the rather easy inclusion of these new technologies in the daily practices of environmentalists. Greens traditionally have always been rather critical towards new technologies, at least until the early 1990s. Although this critical, neo-Luddite attitude towards modern technological systems changed dramatically following the discourse and practices of ecological modernization (see Chapter 3), there is still significant ambivalence towards modern technological systems, for instance, on mobility (cars, planes and even bullet trains), agriculture (GMOs, cloning), and industrial production. From that perspective, the embracing of the ICT system by environmentalists is quite remarkable (cf. Horton, 2004). The non-hierarchical, decentralised, fluid network structure of some parts of the green movement and environmentalism seem to fit rather well the use of the network and nonhierarchical character of the Internet, e-mail systems and mobile communication technologies. Some even speak of an organisational and ideological *Wahlverwandschaft* between NGOs and social movements, on the one hand, and the Internet and ICT, on the other (e.g., Scott and Street, 2001), leading to mutual influences in the development of both.[3] Since the 1990s, national NGOs, and after that also local NGOs, have been quick in seeing the advantages of ICT systems. Environmental NGOs have, for instance, used ICT technologies in their disclosure politics in the 1990s, such as Scorecard (developed by the American Environmental Defense Fund), Factory Watch (developed by Friends of the Earth UK) and *Recht om te weten* (developed by the Dutch Gelderse Milieufederatie).

Whereas, indeed, the spreading of this technology has been impressive among greens, there are also clear backdrops and criticisms. In developed countries, especially women and nonwhite environmentalists have been underrepresented and relatively late in the use of the Internet, both because of limited access (for money and space reasons)

[3] Castells (2004) claims that the cultural revolution of the 1960s and early 1970s, which founded, among others, the environmental movement, was crucial for the open, nonprivileged and nonpatented development and design of the ICT system.

and a lack of skills.[4] With the further penetration of these technologies, this seems to be changing now. In lesser-developed regions access
to the Internet has been a much wider problem for environmentalists
and greens, for instance, in keeping updated with information, campaigns and networks in a globalising green community. The digital
divide, of course, does not stop at the environmentalists' front door.
Although national environmental NGOs in the main urban hubs in
developing countries have been assisted with computer technology and
Internet access by their counterparts in the North and through official
development assistance, the more local organised and oriented NGOs
are still significantly deprived of the global informational flows and
networks. Second, the significant increase of Internet use by environmentalists and NGOs also has drawn major attention of law enforcement agencies and corporations.[5] Their surveillance of the contents of
Web sites and e-mail traffic, the seizing of computers and the national
and international linking of databases of environmentalists (especially
in its more radical forms) might have not been that large and significant when measured against the background of the entire global
green community, but it did make specific sections of the greens para-
noid about the risks of surveillance online and more cautious in using
ICT and the Internet. Foreign hosts, encryptions, computer security,
establishment of mirror sites and resistance against Internet legislation have been some of the actions taken by environmentalists to protect themselves against this vulnerability. "Environmentalists have been
part of a larger constituency who are keen to protect cyberspace as
a public space, free from too much state and corporate dominance"
(Pickerill, 2001: 367). The 9/11 event and its aftermath has certainly
changed the battlefield for the worse, especially for more radical segments of the greens. Recently, organisations, especially those such as

---

[4] Based on her research in the United Kingdom, Pickerill (2000: 15) provides a
useful categorization that illustrates the variety of computer and ICT use among
green activists, ranging from computer sceptics, information gatherers,
connected street activists, desktop activists and online coordinators to a
techno-elite of cyberactivists.
[5] Several large corporations have active Internet managers who constantly
monitor Web sites and engage in cyberspace communications. Several national
intelligence agencies have become very active in cyberspace surveillance with
respect to, for instance, animal liberators, road blocking protestors and radical
environmental activists in general. This dates from before 9/11, but has, of
course, not diminished since then.

Reporters sans Frontières (Reporters without Borders), have taken the lead in defending free access to and free production of Internet sites (see also Chapters 9 and 10). Third, and finally, although the acceptance and glorification of the computer/Internet is widespread among the greens in developed regions, recently a discussion has emerged on the environmental justice effects of the computer technology and ICT system, focusing on the environmental impacts of computer production in peripheral regions, on labour conditions at various production sites in developing countries and the dumping and 'recycling' of especially U.S.[6] electronics waste in the South (cf. Smith, Sonnenfeld and Pellow, 2006). These worries refer back to the original environmental assessment of the Information Society, as reviewed in Chapters 1 and 2, with the difference that now the emphasis is on the environmental side effects of the ICT revolution in *developing* countries, and less so in the North. Although the environmental side effects of Silicon Valley have been known for a while, it has been especially the displacement of the consequences for environment and labour conditions to developing countries and regions – with the fruits to be found in the hubs and nodes of the global networked economy – that triggered recent debates (see Box 8.1).

Consequences of ICT for local environmental organisations

It is remarkable to see how often research on the use of ICT by, and on the consequences of ICT for, social movements take the green movement as an example (especially as the environment has been lacking in the Information Society/Age studies). For example, a volume on the (new) culture and politics of the information age, edited by Webster (2001a), is almost completely on environmental NGOs. Also, the introduction to a more recent volume on cyberprotests by social movements (van de Donk et al., 2004a) starts with the environmental movement as the main example. There seems to be a wide – although unarticulated – feeling that the environmental movement is a useful yardstick to investigate and assess the consequences of ICT and the Internet for

---

[6] This is less true, for instance, for European electronics and computer waste, as a result of more stringent EU regulations on waste (cf. the Directive 2002/96/EC of the European Parliament and of the Council of 27 January 2003 on Waste Electrical and Electronic Equipment).

social activism. But the subsequent assessment of the influence of these new technologies and the related informational practices on the ideologies, strategies and organisational networks of environmental NGOs is far from universal. Broadly, two – not necessarily conflicting – assessments prevail: one focussing on local environmental activism and one on global activism.

To what extent has the Internet undermined, transformed or strengthened conventional local activism and place-based community building via face-to-face interaction and conventional media? Putnam's (2000) social capital thesis argues that the Internet and ICT reinforce conventional community life, face-to-face interactions and place-based networks rather than replacing them. Although the Internet and information technologies clearly widen the opportunities to interact, build networks and form activist communities beyond any place-based restrictions, there are clear evidences that most local environmental groups and organizations use modern information and communication technologies rather to reinforce and strengthen their place-based local relations, instead of moving their attention predominantly to newly created placeless communities. Horton (2004: 742) provides an illustration of this strengthening of local green communities by the combination of virtual and copresent densities of local green networks in analysing how the Internet is used for Swap Shop.[7] ICT and the Internet seem to be strengthening a local green identity as they consolidate rather than transform or undermine existing patterns of local activism/environmentalism, much in line with what is found within new social movement research in general. This is in contrast to, for instance, television technologies, which are known to have caused an erosion of local community life, identity, organisations and activism. The Internet and ICT, enabling two-way communication much more than the TV, have increased a quantitative jump in volume

---

[7] In Lancaster, U.K., activists set up a weekly e-mail bulletin in 2001 to which subscribers could contribute information on the used goods that they either have to offer or want from others. Through this virtual space, green materialities were institutionalised, which was usually done in face-to face interactions. After contact via e-mail, successful transfer of used goods, of course, included face-to-face interactions. In addition, corporal Swap Shop gatherings were organised after the e-mail bulletin proved to be successful. By October 2003, the bulletin had more than two hundred members.

and frequency of communication through its increased easiness, and as such are facilitating new weak ties, strengthening existing ties and building network densities (Diani, 2001). With that, ICT builds on existing ties and relations, rather than creating new ones that did not exist. Virtual networks work best when they are backed by 'real' face-to-face or copresent social linkages in specific localised communities. Computer-mediated communication remains, then, in general restricted to that of an additional communication tool (Pickerill, 2001). Thus, according to these scholars, new information and computer-mediated technologies strengthen existing green cultures and networks, rather than transform or even undermine green politics and activism. That is also what UNESCO (2005) found more generally (beyond the domain of environmental activism), when it investigated the relation between new technologies and political and civic activism: the use of Internet correlates highly with civic and political activism (Figure 8.1).

Although the Internet and ICT have not directly transformed local environmental politics of greens, it has done so in the facilitation and construction of global networks. It is remarkable how swift and suddenly the Internet has become crucial in global environmental activist networks. Some of the major international NGO studies give evidence of that. In his thorough study on transnational activist groups in 1996, Paul Wapner hardly mentions the Internet and ICT as resources of any importance in the struggle of globally operating NGO networks against states and transnational companies. Nor do items as information, transparency and reputational capital belong to his index list of entrances. Similarly, two years later, the much cited study of Keck and Sikkink (1998) on global activist networks pays marginal attention to the role of ICT and the Internet. It is only a few years later, especially following the interpretation and understanding of the Seattle protests in 1999 and its follow-ups, that the Internet and ICT seem to have moved to the centre of social science interpretations of the developments and strategies of a green global civil society (cf. Webster, 2001; Pickerill, 2003; van de Donk et al., 2004b). There are not many other sectors in modern society in which Internet and ICT has made a similar sudden and rapid change in internal organisational modes and external strategies. We now turn to these transnational informational politics and strategies of not just well-organised NGOs, but especially of a more fluid networked global civil society.

3. Transnational spaces for environmental movements

Today, one can hardly envision a thorough analysis and understanding of the anti- or other-globalisation movement without paying close attention to the Internet and computer-mediated communications. Global protests at the Seattle, Genoa and Cancun summits (and at all other summits in the early years of the new millennium) of the leading global economic and political institutions have strongly been enabled, facilitated and strengthened by these new communication and information systems. But also understanding the strategies and networks of more organised, structured and institutionalised global NGO players such as Greenpeace, Friends of the Earth and the World Wide Fund for Nature is no longer possible when the Internet and ICT are not taken into account. Cleaver (1998) and Chesters and Welsh (2005) refer to this ICT centrality in social movements as the 'electronic fabric of struggle'. Although at a local level the Internet does not really increase the ability to mobilize participation in protest events (Pickerill, 2001), arguably this is different internationally. And, perhaps even more important, the preparations by and the feedback to the wider environmentalist constituency – so crucial in constructing a global movement – was built on the new technological systems with their time-less global information flows. The global interlinking of the many local campaigns construct the feeling of being part of a truly global (anti-capitalist, antiglobalisation, antinuclear, anti-GMO, etc.) movement. It is here that working methods, communication patterns, information exchange and/or decision-making structures really have shifted to virtuality, and that corporeal interactions have reduced in importance. In their study on Web sites of NGOs in the antiglobalisation movement van Aelst and Walgrave (2004) give evidence of the role of ICT in global network formation through joint mobilization, Web site linkages, and the development of common ideas, discourses, frames and counterexpertise. Although also in these global networks, face-to-face interactions are essential to build trust, facilitate cooperation and confirm jointness in goals and identities,[8] cyberspace interactions do

---

[8] Global networks, for example, of Greenpeace and Friends of the Earth make ample use of the Internet but also meet face to face regularly to plan and discuss global campaigns, compare national priorities and strategies, and discuss the allocation of resources. These meetings with corporal presence are also essential in building trust and continuing financial solidarity between the resource-rich

more than just confirm weak ties, as we concluded for local, grounded communities. But, indeed, as van de Donk and colleagues (2004b) and others have stressed, the Internet will not easily fully replace traditional forms of protest.

A clear example of these changes is the protests of the anti- or other-globalisationists at the turn of the millennium. On 18 June 1999, several thousand people demonstrated in the City of London in the Carnival against Capital. What could have been seen as one out of many national protest meetings against the major international financial institutions was markedly different from others, according to Scott and Street (2001: 41). It was different because (i) it was one of the most violent demonstrations in London for a long time; (ii) the usual planning arrangements between organisers and the police were not made in advance; (iii) it was coordinated with several simi- lar events in other major cities around the world, coinciding all with the G8 summit in Germany; and (iv) it was organised almost com- pletely through Internet in a decentralised way, without one major formal NGO holding responsibility. According to Scott and Street, this Carnival against Capital seems to be a model for a new form and organisational mode of civil society protest. Although it is much too early to draw final conclusions as to whether or not this model is here to stay (and arguably some of these characteristics – such as the violent character and the absence of planning with authorities – have not really remained dominant in global civil society protesting), the Internet and ICT do seem to have a long-lasting effect on global civil society.

In the fluid networks of global activism, the Internet has become an organisational force shaping the network relations as well as the organisational nodes themselves. The dynamic networking, unpre- dictable traffic patterns and openness of the Internet may give sudden rise to NGO organisations from obscurity to centrality and permits organisations to join and leave networks relatively easily. The rela- tive stability of the national environmental NGO landscape during the 1970s, 1980s and early 1990s is complemented by the end of the second millennium by a much more dynamic and flexible pattern of global NGO configurations, networks and coalitions, as, for instance, shown by Jubilee 2000 and the A16-2000 umbrella organizations. The

and the poorer national Greenpeace subsidiaries (interview Greenpeace campaigner, 2005).

functioning of the World Social Forum,[9] a flexible network of different kinds of NGOs from all over the world who come together regularly at different places, has been possible only as a result of the existence of Internet and ICT. Some, however, have criticised the World Social Forum for relying too much on a political-institutional organisation scheme and too little on a communicational-networked arrangement (cf. Waterman, 2005). These virtual and 'corporal' umbrella platforms bring constantly changing coalitions together for protests or activities, to sometimes melt into thin air directly after these protest and activities. At the same time, the position of the more formalized NGOs participating in those campaigns and manifestations is far from stable and predictable. Sometimes they are the main constituents and organisers; at other times, these NGOs step in an existing fluid network when it is already functioning, or they step out of such global networks just before major events give these networks or fluids global significance. In addition, as Bennett (2004) correctly argues, the virtual networks tend to be less strict in ideology and accommodate a diversity of groups and activists under rather loosely defined common denominators as child labour, sweat shops, fair trade labels or corporate globalism.

In one of the few empirical surveys on the role of the Internet and ICT (or computer-mediated communication, CMD) in global activism, della Porte and Mosca (2005) interviewed protesters at the anti-G8 protest in Genoa (July 2001) and participants in the European Social Forum ESF in Florence (November 2002). They were especially interested in the contribution of the Internet to organising, protesting, identity formation and cognitive functions. With respect to the use of Internet in organising protests, they found that the Web and CMC had indeed become the key organisational devices, as they lowered costs of mobilisation, facilitated logistics and eased communicating the content of the protests. The protesters who mostly used the Internet were well educated, young and had already experienced computer technologies in their organisations (which was not as common in 2001 as it was in 2007). Protesting via the Web took various forms, online petitions being most popular among the participants (66 percent), followed by creating Web sites with names close to the organizations they aimed to address,[10] netstrikes and mail

---

[9] The WSF is the alternative of NGOs to the World Economic Forum, where the political and economic elite meet every other year.
[10] Such as http://www.worldbunk.org; http://whirledbank.org; http://www.gatt.org; http//:www.genoa-g8.org; and http://www.seattlewto.org.

Table 8.1. *Correlations (Kendall's Tau B) between online and offline forms of actions by ESF activists*

Offline forms of action (dummy variables)	Online forms of action	
	Petition	Netstrike
Taking part in past demonstration	0.05*	0.04
Squatting	0.04	0.16**
Taking part in elections	0.07**	−0.14**
Taking part in sit-ins	0.06*	0.14**
Perpetrating violence against property	−0.02	0.13**
Taking part in boycotts	0.15**	0.12**
Handing out leaflets	0.07**	0.11**
Taking part in strike action	0.02	0.08**
Attending political meetings	0.06*	0.06*
Occupying school/university	−0.03	0.06*
Canvassing for a political party	0.05	−0.05
Signing petition/referendums	0.18**	−0.04
Party activism	0.03	0.02

* = significant at 0.05 level; ** = significant at 0.01 level.
Source: From della Porte and Mosca, 2005: 178.

bombing.[11] Although it often has been claimed that online activism replaces offline activism, della Porte and Mosca (2005) found rather strong relations between online petitioning and netstriking and several offline forms of protest (cf. Table 8.1). With respect to cognitive information dissemination, the Internet proved to be widely used among protestors both for their own information as for spreading their ideas. Although more open and easy accessible, the Internet also proved to possess new hierarchies and gatekeepers, and knowledge of the rules of search machines is essential for prevalence on the Internet. Most protestors also were well aware of the lack of verification and the highly temporary nature of information on the Web, compared to conventional media. And, finally, the transfer "from desktop to television screen" (Bennett, 2004) proved to be a hard one, as the contents of the Genoa and ESF protests were only marginally covered by the

[11] Netstrikes consist of a large number of protestors connecting simultaneously to the same domain at a prearranged time. This prevents regular users from accessing the domain. Mail-bombing consists of sending an overload of e-mails to a Web site until it gets jammed.

conventional media (compared to news of the official summit). Conventional media seems to be more interested in covering law and order conflicts of these protests than in substantive issues, although these protests were successful in setting the dark side of globalisation on the agenda.

Quite remarkably, environmental activists and NGOs hardly use the Internet and ICT to block or frustrate informational processes of state authorities, transnational companies or other targets of environmentally unfriendly behaviour. Netstrikes, e-mail bombing, computer hacking or other subversive actions using the Internet have been used to a very limited extent by green activists, although there has always been a core of techno-elite cyberactivists among the greens that are, in principle, able to (and sometimes do) develop such strategies. Only incidentally – on the peripheries of the environmental movement – have such destructive strategies on the Internet been applied.

The Internet also provides global activists with a considerable degree of information, communication and media independence, compared to the conventional information and communication technologies, media and networks (cf. Scott and Street, 2001: 46). This is especially crucial because environmental NGOs and civil society rely so strongly on informational resources and power in their confrontation with the powers of state and capital. Both on collecting and on disseminating information weblogs, Internet search engines, e-zones, listservs and their own media networks (such as Indymedia; see later in this chapter) have greatly stimulated the effectiveness, global operations and independence of the global greens. In addition, and most vitally, these new media outlets also have been discovered by the conventional media as a source of information and news. There is increasing evidence that the underground protest information of activist Web sites and e-mails bypass conventional 'gatekeepers' (journalists and newsworkers) and make it directly into the mass media, enabling the destruction of the reputation of corporations or the downfall of ministers or governments within a few months. In the Information Age, reputational capital of the private sector, often crystallised in logos and brands (cf. Klein, 2000) has been the key target of both green activists and private corporations, as Nike, Shell, Exxon, Microsoft and numerous other companies have witnessed.[12]

[12] The case of Nike shows how significant information is nowadays. At the turn of the millennium, mainly in the United States, Nike was one of the main

However, the global green networks have not really succeeded in using the Internet and ICT to develop a full alternative to the existing mainstream news and media. Most information politics by global greens are fragmented, focus on temporary campaigns rather than systematic issue coverage over longer time horizons, show large diversity, have limited reach in terms of audience and remain obscure. There are, of course, a significant number of Web sites and organisations that specialise in diffusing information of (environmental) NGOs by the Internet. OneWorld, for instance, claims to be the online media gateway that most effectively informs a global audience about issues of sustainable development, via a variety of thematic portals.[13] But the only serious challenge to the conventional media might be Indymedia. Indymedia[14] was founded in Seattle in 1999, during the protests against the World Trade Organisation (WTO). It was designed to offer new, alternative news sources using the Internet, which challenged conventional media in various characteristics and, at the same time, constructed affinity, relations and identities to various parts of the more left-wing civil society, engaged in anticapitalist and anticorporate globalisation. Although Indymedia continues to evolve, its initial ideas, codes and core characteristics have sufficiently stabilised.[15] "Indymedia has become the fastest growing, international, alternative media network in the world, mushrooming into dozens of physical and virtual sites that span six continents" (Nogueira, 2002: 294).[16] Although Indymedia is not so much an alternative for the large media conglomerates that dominate media production, it does challenge these oligarchies that

targets for informational campaigns by NGOs, because of its poor labour and environmental conditions, especially in its Asian factories. In May 2005, Nike started to publish name and addresses of all its seven hundred supplying factories on the Internet, in a unique move to full transparency and to safeguard its reputation. Although its 2005 Corporate Responsibility Report shows still significant problems with respect to labour conditions, the situation in many of Nike's (supplying) factories did improve.

[13] See http://www.oneworld.net.

[14] See http://www.indymedia.org; see Morris, 2004.

[15] It has grown from one outlet to more than one hundred in three years (Bennett, 2004).

[16] Inter Press Services (IPS) is another news agency that challenges the dominant commercial international news agencies, by aligning more strongly with NGOs, remaining independent from advertisements, focussing strongly on cross-cultural communications and giving more access to developing countries. With a network of journalists in more than one hundred countries, and editorial offices in most regions, its structure comes closer to the main transnational media conglomerates than Indymedia (cf. Chapter 10).

determine what is newsworthy, by a more fluid understanding of truth and a more dispersed production of news.[17] Against the hierarchically, formalised, privately owned and closed structure of conventional media conglomerates, Indymedia is more mobile and decentred, having multi-ownership, an open structure, free access, 'copyleft' and limited filters and editorial interventions that mediate between the decentred producers of 'news' and the final publication. It can thus be seen as a practice in participatory democracy, although computer access and technical skill, certain social and cultural barriers (language, poor representation of indigenous groups) and geographical concentration to Western countries prevent the materialization of ideal-typical full democracy (Pickerill, 2003b).[18] Overall, Indymedia glues various (leftist) protest movements together globally, constructing perhaps not directly a global community but certainly new ties and a loose sense of collective identity among groups that were previously separated in place, time and issues. It has proved to become essential for a global, active civil society in an age where information is crucial.[19]

4. New strategies, new alliances

Our analysis of environmentalists in informational governance has up until now focussed mainly on the Internet and ICT. There is, however, also a category of civil society innovations in informational governance that is linked less to ICT and the Internet. The growing importance of information in environmental governance, and the importance of

[17] Wikipedia, the Internet encyclopaedia that is constructed and continuously renewed in a decentralised way by all its users, has a similar structure. Although a comparison illustrated that the correctness of Wikipedia can compete with the commercial Encyclopaedia Britannica, increasingly it has become clear that Wikipedia is strategically used for private interests and has the danger to turn into an informational battleground. In 2005, it was discovered that climate change sceptics were systematically reversing all information in Wikipedia that pointed at the seriousness and dangers of climate change. In 2006, staff of U.S. congressmen 'cleaned' more than one thousand Wikipedia articles (often biographies of politicians) to make them more advantageous to their politicians. These and other problems led to calls and initiatives to leave production of Wikipedia to a more limited group.

[18] The development of decentralisation and regionalisation of the Indymedia network is one reaction to the (perceived) cultural dominance of the United States. Discussions on language continue at Indymedia.

[19] Not surprisingly, there have been several attempts to limit the influence of Indymedia by state censorship and security and intelligence agencies (cf. Pickerill, 2003b).

legitimacy, trust transparency and accountability that come along with that, have strengthened the position of environmental NGOs in environmental governance. These organisations are crucial managers of trust and legitimacy, and through that they fortified their position in the environmental governance arena. In Chapters 5, 6 and 7, we have seen several examples of how civil society actors manage to enhance their position and influence in environmental governance, for instance, in labelling and certification practices and arrangements, in disclosure programs, in monitoring and verification. Some of these innovations make use of and run via ICT and the Internet, but this is often not an essential precondition.

These new power balances give also rise to new strategic alliances between civil society actors and political and economic sectors. In the early days of environmentalism, in the 1970s and 1980s, environmental NGOs were mainly positioned at the periphery of environmental policies and reforms, only able to criticize and disclose the unwillingness and failures of political and economic powers. As emphasised in the ecological modernisation literature and ideas, we have witnessed a repositioning of mainstream environmentalism, more towards the centre of economic and political decision making.[20] With the growing importance of informational governance, this repositioning is even further strengthened. Environmental NGOs are able to use their 'informational powers' of trust-generators, legitimacy providers and transparency watchdogs to build strategic alliances with the powers that be. Representatives of civil society are increasingly included in major decision-making structures of national institutions; they sit regularly around the table with major international bodies such as the World Bank, the IMF, the OECD or the World Economic Forum; they set up institutional arrangements with national and transnational companies around labelling schemes, environmental reporting, codes of conduct and environmental disclosure; and intelligent national and transnational companies involve NGOs in their assessment procedures for new products and investments or engage them in identifying or even promoting green niche markets. These positions are more and more institutionalised and/or legally codified, for instance, through theÅrhus convention and the inclusion of that in national and EU legislation.

---

[20] On an earlier occasion (Mol, 2000), I have elaborated on the repositioning of the environmental movement following ecological modernisation developments and discourses.

It goes without saying that these developments go together with internal debates within the environmental movement on strategies and goals between radical and reformist factions, and to some extent also between northern (or Western) large and well-resourced environmental organizations and smaller NGOs from the South (or northern ones identifying themselves with those from the South). Especially around the symbolic flagships of global capitalism the minds and strategies of NGOs split, for example, on environmental NGO cooperation with the IMF, Shell, McDonald's or the World Bank. How can environmental NGOs in the Netherlands cooperate with Shell on solar power, when Shell causes so much trouble for the Ogoni in Nigeria? How can international NGOs cooperate in developing environmental conditionalities for World Bank loans, when many of the World Bank projects have disastrous consequences on localities and for indigenous people in developing countries? It is these kinds of dilemmas that a global civil society is facing. Although the powerful institutional actors in the space of flows have often major incentives for allowing environmental NGOs in their global networks, resistance against such inclusion will remain among the more radical grassroots factions of the environmental movement. In that sense, the environmental movement is no longer just a place-based resistance movement against the global networks that make up the space of flows, as Castells (1997a) suggested. Nor are environmental NGOs completely part of the space of flows. Increasingly, environmental NGOs have become one of the mediating corridors where the space of place logics connects with the space of flows.

5. Legitimatory capital at risk

The global environmental movement is the originator, the advocate and the final judge of global environmental norms and values. From the birth of the modern environmental movement onwards, the environmental spokespersons of civil society have disclosed and judged on environmental norm violations, either the legally set norms or those prevailing in civil society. The production, distribution and strategic use of information have always been the key power resource in this process. But that has only been effective through the movement's legitimatory capital. It is, as Ulrich Beck (2004: 239) rightly remarks, the linkage of information with legitimation that has provided the environmental movement with the ability to become a powerful actor in informational governance, even in those days when informational governance was still underdeveloped. With the growing centrality of information, informational processes and informational resources in environmental politics and governance the (trans)national environmental movement witnesses an improvement of their power position vis-à-vis their main counterparts in the political and economic sectors. But that continues only as long as their legitimatory capital is not put at risk.

Legitimatory capital is nonconvertible with economic or cultural capital; legitimation cannot be bought, nor can it be transferred easily from cultural capital (and, hence, it is different from the forms of capital originally defined by Bourdieu). As much as the other forms of capital, legitimatory capital is not an absolute quantity but, rather, a relation, defined between different groups. It is the fragility of legitimation of global corporate power, especially in the field of the environment, which provides the environmental movement with significant legitimatory capital. But legitimatory capital does not just fall on an actor, it is not pregiven, and therefore it needs to be actively maintained. Legitimation is gained through long-term accumulated experiences of credible behaviour; continuous release of reliable information; and measured against accepted standards, norms and values of civil society. Steady high levels of legitimatory capital of environmental movements, especially when compared to public authorities and economic actors, have been illustrated by European survey research over the past one and a half decades (Table 8.2).

Through their alliances with other actors and sectors, and through their strategic use and disclosure of information in environmental politics, legitimatory capital of environmental NGOs can be at risk. When information is used to dramatise events and practices in order to achieve certain goals, and when coalitions are built that are considered only of strategic, short-term and particular gains without taking the partner's overall performance into account, then environmental NGOs might be considered just to be one out of many parties with their own specific interests, rather than a movement looking after and safeguarding universal norms and values. This was exactly the case in Greenpeace's misleading interpretation of the contents of Shell's oil platform Brent Spar in 1995.

The multiple sources of information the reliability of information and the lack of an authority to shift true from misleading information also may interfere with legitimatory capital of NGOs. As the environmental movement – more than most other social movements – draws

Table 8.2. *Trustworthy sources of environmental information in EU 15 (in percentage of respondents)*[21]

	1992***	1995	1999	2002	2004
Environmental protection associations	63%	63%	51%	48%	42%
Scientists	50%	51%	37%	32%	35%
Television*	23%	24%	27%	18%	26%
Green political parties	–	–	11%	10%	12%
EU	–	–	–	13%	10%
National governments**	12%	13%	9%	12%	11%
Companies	4%	2%	2%	1%	2%

* In 1992, 1995 and 1999 the category was media.
** For 1992, 1995 and 1999 the category was public authorities.
*** EU 12.
Source: Eurobarometer.

heavily on science (or what Castells [1997a: 123] calls a science-based movement), it has to cope with the structural uncertainties in information and the disenchantment with science that characterises global modernity. Wright (2004), for instance, points to the fact that problems faced by conventional governmental actors in information management are equally relevant for environmental NGOs and activists. Excess of the volume of information, uncertainty on the reliability of information and knowledge, and chaos in the organization of information flows seem to challenge the configurations of power in the internal and external environment of NGOs. This is illustrated by Grignou and Patou's (2004) analysis of Internet use by ATTAC.[22]

[21] Sources: European Commission (2005), *The Attitudes of European Citizens towards Environment*, Special Eurobarometer 217, Brussels: DG Press and Communication; European Commission (2002), "The Attitudes of European Citizens towards Environment", *Official Journal of the European Communities* L242/1; European Commission (1999), *Environnement? Ce que les Européens en pensent*, Eurobarometer 51.1, Brussels: DG XI/DG X; European Commission (1995), *Europeans and the Environment*, Eurobarometer 43.1 bis, Brussels: DG XI; European Commission (1992), *Europeans and the Environment*, Eurobarometer 37.0, Brussels: DG XI. The questions and categories have not remained exactly the same throughout the years.

[22] ATTAC stands for Association for the Taxation of Financial Transactions for the Aid of Citizens, was established in 1998 in France, and grew rapidly to an international movement.

Especially in an Information Age, the countervailing powers challenging the environmental movement will point their arrows primarily
at the legitimatory capital of environmental NGOs. In their informational and conventional struggles on the environment, questioning the
reliability, truthfulness and credibility – or, to be short, the legitimacy –
of environmental NGOs is one of the most frequent strategies followed
by companies and authorities who challenge environmental NGOs,
that is, when more classical resources, such as limiting press freedom
and restricting the freedom of establishing NGOs, are no longer possible (see Chapter 10). As much as in the Information Age reputational
capital becomes the Achilles' heel of the transnational companies, it is
the same with legitimatory capital for the environmental movement.

6. Conclusion

The Internet and ICT-mediated communications have made a significant difference in the operations and structure of a global civil
society, which aims to defend environmental sustainability. To some
extent, parallels in innovation can be drawn with what happened in
the global networked economy, where environmental challenges were
addressed via new organizational schemes, new (marketing) techniques
and strategies and new forms of legitimation.

The virtual space of the Internet has become a crucial constitutive
component of a local green community, through the constant multici-
plicity of e-mail bonds. This happens in addition to a more dispersed
green social network, stretching beyond the local. Today, there is a
more complex interweaving of activists' virtual and corporeal social-
ities and geographies (Dodge and Kitchen, 2001; Miller and Slater,
2000). Although this all clearly tends to strengthen the activists' green
identities, at the same time it transforms the conventional place-based
local green identities to more global ones. This means that the Internet
and ICT cannot be ignored in understanding today's environmentalism,
both as a reinforcement and constitutive power of local green communities, and as a formative power of building green networks beyond the
local and the national. Although the Internet and ICT might not have
a major transformative impact on the internal organisation and functioning of local green communities, it has been a key factor in changing
the opportunities, strategies, organisational modes and countervailing
powers of the green civil society.

In that sense, Castells's (1996/1997) analysis of the environmental movement as a place-based, grounded countermovement that makes full use of modern information technologies to address the consequence of processes located in and organised through the space of flows is only partly adequate. More than Castells imagined one decade ago, a significant part of the green movement engages with the space of flows; and the structure, dynamics and strategies of that part of the movement have changed through processes of globalisation and informationalisation. The evaluation of that development differs within the environmental movement itself but also by outside commentators: from condemning a movement being incorporated in global corporate capitalism, to celebrating a strategic innovation following a new time frame in which full use is made of the informational powers of a globalised civil society.

If there is a central space where these informational powers of the (globalised) environmental movement clash with the informational (and other) powers of economic and political agents, it will be the media. With informational politics and governance becoming more central in environmental controversies and struggles, we can expect the media (i) to become more and more the location of environmental controversies, (ii) to be used as a crucial resource in environmental controversies and (iii) to be transformed through such informational politics. The next chapter focuses on these media hypotheses.

9 | Media monopolies, digital democracy, cultural clashes

1. A New World Information and Communication Order?

In the 1960s and 1970s, following the decolonisation and debates on how the structure of world capitalism affected the international order, the notion of a New International Economic Order was launched, as a critique against the distortions and inequalities of the existing international economic order. Following similar lines of analysis, the distortions of and inequalities in the world's international news and information structure were criticised in the 1970s, especially for the concentration of media power within a few mighty news agencies. This resulted in a call for a New World Information and Communication Order (NWICO; cf. Gaber and Willson, 2005; McPhail, 2006: 241–66). According to the NWICO proponents, (i) the conventional, existing information order had (and still has) a highly unjust and inequitable balance in the flow and content of information between OECD countries (and the major nonstate actors related to them) and the South; (ii) there should be a right for countries to self-determination and sovereignty of domestic communication and information (in- and out-) flows and policies; and (iii) internationally, a two-way information flow should more accurately reflect the aspiration, activities and interests of developing countries, rather than that dominant northern media conglomerates create, reproduce and transmit stories on the South only as they relate to famines, wars and disasters. The aim of NWICO was to restructure the system of media, information and telecommunication, so that developing countries gained larger influence and self-determination not only over their media and information flows, but also over their economic, cultural and political systems. This analysis fit well within more general ideas that circulated in the Information Society debate of the 1970s (see Chapter 2). But diverging from the Information Society scholars was one of the solutions proposed by the NWICO analysts to deal with the media and information distortions and inequalities: the idea that states had the right to control the flow of information moving in and out of countries. States – of which not all were that democratic – would have the right to decide which information and news should reach their citizens and what information and news from their country would be available for the outside world. Under the banner of 'developmental journalism', governments were given the justification to censor the flow of incoming and outgoing information and news and to stop journalists and news agencies that did not contribute to the state's developmental process. Although it was aimed to limit the power of northern media conglomerates and strengthen the media position of the South, this would, of course, seriously hamper the media work of NGOs and independent journalists, unless they would act and communicate fully in line with the official government policy. This NWICO solution was most strongly debated in UNESCO from 1970 onwards, culminating in the withdrawal of the United States, the United Kingdom and Singapore from UNESCO in the mid-1980s. Although freedom of expression arguments were one major line of criticism against these NWICO proposals, in the United States and the United Kingdom more right-wing think-tanks and media interests pushed the Reagan and Thatcher governments to leave UNESCO. Although the analysis of NWICO on media monopolies, distortion and northern dominance today remains largely intact and is, for instance, theoretically reflected in electronic colonialism theory (McPhail, 2006), its drastic solution of state control on information is no longer actively supported by many.[1]

Since the 1970s, the importance of the media has not only increased but also has become more complex, making analyses and assessments of the role and position of the media even more pressing. The debate on what we will call the 'old' or conventional media (television, radio, newspapers) recently has been renewed and complicated with the emergence of the 'new' media of cyberspace, weblogs, personal computers and the Internet. Positions and arguments with respect to the old media are not always the same and valid for the new media, as we will elaborate later in this chapter. But, at the same time, one can wonder how

---

[1] There are, of course, constant attempts by various states to do so. Following 9/11, state influence on the media has been growing in the United States. See also the example of the fifteen states that try to increase state power over the Internet (Chapter 10).

long this distinction between old and new media will remain adequate
and useful in media analysis, as the borders between old and new are
increasingly blurred when weblogs and newspapers get increasingly
intermingled and television becomes more and more interactive and
related to the PC. But, still, it is relevant to analyse what will be the
consequences of the emergence of the new media around Internet for
the mediascape. Has the emergence of this new media given birth to a
new world media order, or will the new media in the end be encapsu-
lated in the 'conventional' media and information structure? George
Barnett (2004, as quoted in McPhail, 2006), for instance, draws our
attention to some of the similarities between the Internet and the old
media, such as the central and dominant position of the United States
as the nucleus of Internet traffic and the fact that direct communications between informational peripheries are rare, as most communications go via the United States or other information hubs. Boyd-Barrett's
(2004) analysis of the U.S. hegemony in global cyberspace also follows
this line of analysis.

Our purpose here is not to answer these questions as that would
require a full-fledged media analysis and evaluation. We will restrict
ourselves to the role of the media in the emergence of informational
governance of the environment. We will start our analysis with a review
of what I would label conventional environmental studies on the media.
In concluding that most of these approaches have been useful but limited, we extent our analyses to the transformations of the media and
the consequences of that for informational governance. We start that
analysis by a more fundamental exploration of what the media does to
information and experiences, to be followed by an analysis of the major
media transformations that can be witnessed during the past decade
or so. Subsequently, we will analyse what this means for informational
governance of the environment and the shift of environmental battles
to the media.

2. Mediated environment

Environmental arguments and positions have never played a too strong
role in debates on the media, neither in the debate on the NWICO
nor in other media debates. Most of the environmental studies and
research on the media have been much more focused. Traditionally,
environmental research on the media has strongly concentrated on

media production and products (see, for instance, several chapters in Hansen, 1993b; Lacey and Longman, 1993; Dispensa and Brulle, 2003; Hannigan, 2006).

Anders Hansen's (1993b) edited volume on environment and the mass media gives a good overview of how environmental social scientists have studied the mass media. Three major themes usually stand out to understand how and when environmental issues have and have not been covered and reported on by the mass media and how environmental media coverage has been 'received' by various audiences. In the production of environmental news and communication interest has been on the ups and downs in news coverage, the role of various newsmakers and gatekeepers in press coverage, and the various strategies and roles of environmental NGOs, state agencies, scientists and polluters to get their stories, views and framing across the mass media. A strong emphasis has been on the social construction and power play of environmental media messages, claiming that only through social constructivist perspectives we can understand the major discrepancies between environmental realities and media coverage. The second major theme focuses on the actual framing of environmental problems and solutions in the various mass media. Content analysis, discourse analysis, social constructivism and critical theory are some of the often used approaches to – qualitatively and quantitatively – study environmental coverage on television, radio and in newspapers, often strongly in a historical perspective through longitudinal studies. How, how often, how framed, in which media and with what timescape are environmental issues mediatised? Third, mass media studies have looked into the ways in which the mass media have changed public understandings and values of the environment. Survey research, ethnographical studies and qualitative case studies have tried to understand the influence of the mass media in society's appreciation of environmental goods, but also regarding the mass media influence in environmental policies, politics, controversies and conflicts. This includes sensibilities on the relevance of the context in which these media messages are transmitted and 'received'.

These research lines have been extremely useful, providing rich – also quantitative – insight in "the considerable extent to which environmental news is socially constructed" (Hannigan, 2006: 92). Most of these studies had a critical nature. These studies often blamed newsworkers and journalists for their superficial, simple, short and limited interest and reporting on complicated, long-running and crucial environmental problems. They also criticised the strong representation – or even domination – of the political and economic powers in the framing of and access to the media when environmental issues are concerned. And these studies took notice of the near absence of environmental sensitivities in the main – nonenvironmental – news reporting and news consumption. Notwithstanding these critical lines, if one has to draw one conclusion out of all these media studies on the environment up to the late 1990s, it should be that the environment became an established part of everyday journalism and media reporting, without one single overarching and dominant environmental perspective, framing or discourse.[2] In that sense, from an environmental point of view, the media matured. But, in addition, the media also transformed – along three lines – in a period that has been labelled the Information Age, and these changes have been reflected less strongly in environmental media studies until now.

First, with the growing importance of information in various domains, including environmental protection, struggles and controversies have relocated strongly to the media. Second, with globalisation and privatisation, the nation-state container becomes less and less relevant for analysing and understanding – also the environmental dimensions of – the media. And, finally, environmental media studies have to take the emerging new technologies – or, rather, new media, – into account. The Internet, cyberspace, new telecommunication technologies and satellites (the new media) are reforming the 'old' media – environment interactions. The consequence is that the media have to be analysed increasingly as a complex system, a mediascape, in which products, production and reception are part of a structured but constantly changing global networked system. In global mediatized information flows through these networks, simple schemes of clearly separated actor roles (producers, transmitters and consumers/recipients), and state governance are no longer adequate.

While most environmental scholars would easily agree on the growing importance of and complexities in various media in most social

---

[2] Hannigan (2006: 89–92) provides an interesting analysis of quite contrasting tendencies that can be witnessed in the mainstreaming of environment in the media, such as scientific objectivism vis-à-vis extreme subjectivism, and greening business and economics versus apocalyptic environmental reporting. In that sense, the media do reflect a variety of societal ideologies and frames.

and political processes, their analysis and assessment of that differs (as they did in the former paradigm/era). The environmental scholar Bill McKibben (1993) is most strong in blaming the media for the age of 'missing information', where all kinds of conventional learning, communication and information exchange processes have been replaced by an inferior substitute: the global media with its commercials and poor informational substance. Other scholars join this criticism by blaming the commercialisation and capitalist mode of media production for the poor informational qualities of the current media. In his environmental criticism of what he labels electronic capitalism Robert Paehlke (2003: 28) is quite straightforward:

Electronic media are a dominant industry in their own right and the principal means by which all other products are branded and sold globally. The ownership of communication capacity is increasingly centralized within large private corporations and is a rapidly growing component of wealthy economies. Global competitiveness, communicated pervasively through the media, threatens to become a universal core sociopolitical value.

But at the same time others are more positive or even enthusiast about the time-space bridging properties of the media (cf. McLuhman and Fiore, 1967; Meyrowitz, 1985), and also on the role these media play in environmental learning processes, in communicating and disseminating environmental information and knowledge across great distances, and in getting environmental messages across to a wide ignorant and mostly indifferent audience in no time with limited effort. And also Robert Paehlke (2003) is less critical of the 'new' media, the Internet, while condemning the 'old', capitalist media.

Before turning our attention to the major changes of the media during the past decade and what this means for informational governance on the environment, we elaborate more theoretically on mediated information.

3. Media and mediated information

For most people globalisation is not felt through their travelling but rather when they stay at home, through the information and communications that come to them via various media. It is what Giddens (1991) calls the centrality of mediated experience that distinguishes late modernity from its earlier time periods. Mediated experience is then the "involvement of temporally/spatially distant influences with human sensory experience" (Giddons, 1991: 243). Mediation thus bridges time and space in communication, through modern communication technologies and through the mass media. Media theorists such as Meyrowitz (1985) have elaborated on the decreased significance of physical presence in the experiences of people, as electronic media have altered the significance of both time and place for social interaction and communication. Face-to-face presence is often no longer relevant in communications through the mediated deterritorialisation of experiences.

But is experience the same as mediated experience? Tomlinson (1999) has questioned the neutrality of the medium in passing experiences through space and time. The medium is not just a technological mean that delivers experiences with high speed, effectively and easily. We should also question to what extent the medium intervenes in the interaction and experience itself, when information is communicated. In fact, the development of the modern media can be seen as a constant attempt to diminish and minimise the intervention of the medium in the communicated experience. Immediacy, in the sense of dissolution of the medium itself, is of course never possible, and in that sense there is a fundamental, qualitative difference between mediated experience and direct face-to-face communication. Tomlinson follows Thompson (1995) in distinguishing three modes of interaction. Face-to-face interaction has a dialogical form with a multiplicity of symbolic cues made possible by a shared spatial and temporal reference system. Second, mediated interaction (letters, telephone, fax, telegraph) is also dialogical, but – by use of a technical medium – allows communication between persons distant in place and/or time. But through that, it lacks common references of copresence and has thus a more narrow range of symbolic cues. Third, mediated quasi-interaction (the mass media) is essentially a one-way (monological, thus quasi-) communication meant for a range of recipients (rather than one). It is still interaction (and not just one-way information transfer), as the recipients are active in their interpretation and meaning construction. The first two categories do not really deterritorialise experiences and interactions in ways that disturbs local cultural and moral narratives, as they place experiences in a meaningful wider context. The latter category of mediated experience potentially can and often does disturb local cultural and moral narratives, as with the technological expanding capacity of the media, news coverage of distant events on television becomes one-dimensional visualisations and as such seems to shrink.

It is here that much of the criticism of television comes together. Television does not succeed in closing cultural and moral distance and thus does not manage to get us sensory-'involved' or engaged in distance practices and events. It is the medium of television that inhibits the closing of distance, and rather works towards alienation, cutting us off from distant events and insulating us from cultural encounters, because the TV screen reduces all experiences to the same one-dimensional visual qualities. It is via these lines that Bauman (1992) criticizes the failed materialization of McLuhman's global village, getting again very close to a polarization between a celebration of direct experience and condemning mass-mediated interaction. Others have less polarized assessments. Following Castells (1996), Young (1990) and Massey (1994), Tomlinson (1999) stresses that this qualitative distinction should not be interpreted as a pure, transparent, simple and morally superior form of face-to-face communicative interaction versus a shortfalling, indirect, symbolized and mediatized interaction. Similar to the idea that there is no superiority of small-scale political communities over large-scale mass societies, there is also no inherent preference of face-to-face versus faceless interaction. Or, as Castells (1996: 373) puts it: "In a sense, all reality is virtually perceived".

Tomlinson (1999) gives two principal reasons for the poor record of television in closing moral distance. First, the intrinsic nature of the medium, the monological character of mass media, means that engagements with mass-mediated events are different from other media. These events are not within the reach of the recipients. Our inability to intervene in the experiences and events many miles away, which are remote from our daily lifeworld and over which we have no control, means that we experience ourselves as being insulated from them. Second, one has to recognize that telemediated experience is situated in a total flow of everyday lived experiences. A vastly growing number of experiences, claims and moral demands comes together with the deterritorialisation of experiences especially via – but not only – the mass media. The 'perceived relevance structure' of experiences (Thompson, 1995) conceptualises that individuals make selections and priorities within the flow of information and experiences. So the nonengagement of telemediated experiences cannot just be blamed on the features of television or mass

media, but should be seen against the background of an information overflow, in which deterritorialised experiences overall have less
'power' and influence than place-bounded experiences.

The Internet, e-mail systems and cyberspace divert from some of the
cultural and moral characteristics of television, as engagements can be
much more direct and the dialogical structure is stronger. At the same
time, the Internet adds up to the information overflow and has similar
features of deterritorialised experience as television. But through its
dialogical structure, the new media is – more often than, for instance,
television – able to support and strengthen face-to-face, place-bounded
and co-presence information exchanges, as we saw with respect to the
use of ICT by environmental NGOs. Consequently, until now the new
media is generally believed to do better in closing moral and cultural
distance than television. In the future, this will depend on the further
development of this new technological paradigm and the way it is integrated with the conventional media. That will determine the further
impact of cyberspace and the Internet on individual experiences, interactions and interventions, as well as on social systems.

4. The Fourth Estate in transition

There are three major developments that have changed the landscape
of media, mediated information, and news dramatically during the past
two decades. First, whereas for a long time the media was contained in
the nation-state, there is an increasing tendency towards global media.
It is only recently that media such as newspapers, television broadcasting and radio have started to address audiences beyond the nation-state
of their origin. Most television stations were national, even for a long
time publicly run and state-owned. The so-called public sphere, in
which the media functioned and which was also structured to a significant extent by the media, was a national public sphere. That has started
to change dramatically during the past two decades. Satellite television,
the Internet and even newspapers have widened their audience and
thus their profile (cf. Schuler and Day, 2004). Although it might yet be
one step too far to speak of a global public sphere with global media
(as Sparks, 2005, correctly notices[3]), the containment of the media in

---

[3] Sparks (2005) points at the limited audiences of truly global satellite TV
  (e.g. CNN, BBC World Service); the difference that the global media brings in

the nation-state system is no longer a useful perspective to understand their functioning and dynamics. Or, in Appadurai's words: the mediascape is no longer a national mediascape, no matter how relevant the state still is in political constraints, regulations, competition rules and broadcasting rights. Although no longer contained in the nation-state, of course there are huge differences with regard to access to these global media throughout the world, as Figures 9.1 and 9.2 illustrate.

Second, the globalisation of the media industry further enhanced a process that was already on its way: the privatisation of once public media and the subsequent concentration of ownership in the hands of a few monopolists. There are a few multimedia transnationals that have a major influence on the global mediascape, Time-Warner, Disney, Viacom, Bertelsmann, News Corporation and Sony among the leaders. This is not only true for television (where, for instance, CNN, Fox News and the BBC are currently leading global news bringers, often strongly influencing whether politicians act or not[4]), but also for newspapers, magazines and, increasingly, book publishers. Only cable television and the Internet seem to have avoided such tendencies of concentration and monopolisation. This makes news increasingly financially driven and controlled. According to Tumber (2001), the business of news organisation and the product are now joined, greatly affecting the substantial part of news, as well as the positioning of news among all other kind and forms of information (advertisements, scandals, intrusion and the like). The commercialisation of news has enhanced significantly over the past two decades. At the same time, especially in countries such as the United States and the United Kingdom, the role of public news agencies and media is shrinking. In parts of mainland Europe, with a different media culture and historic background,

publishing and broadcasting in different localities around the globe; the language barriers; and the role of states in constraining a truly free global media. In addition, one also should point at the monopolisation of these global media. Also the Internet, with its limited access through the digital divide and political constraints in various states, does not qualify for a truly global public sphere.

[4] See, for instance, Robinson (2002) on the so-called CNN effect, which is especially felt strongly in the United States. It refers to the effect that when an item of foreign development is covered by CNN, it becomes a primary concern for the audience and it forces (the U.S.) government to act. CNN is agenda-setting for U.S. foreign policy. "In terms of foreign affairs, if it is not on CNN, it does not exist" (McPhail, 2006: 157). That is, of course, also true for foreign or global environmental issues.

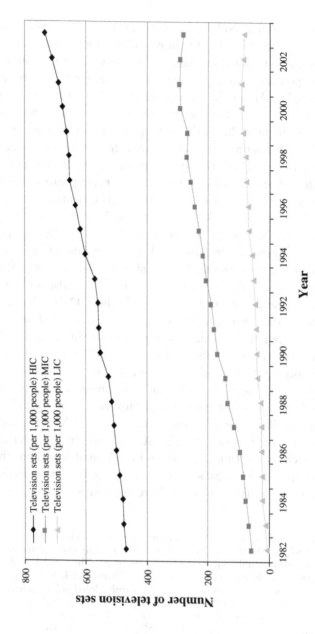

Figure 9.1. Television sets per one thousand people for high income (HIC), middle income (MIC), and low income (LIC) countries, 1982–2003 (*source:* World Development Indicators).

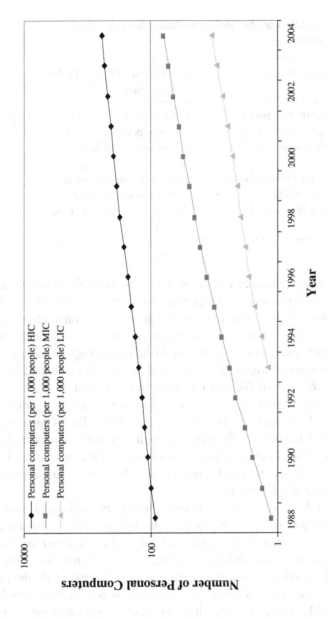

Figure 9.2. Personal computers per one thousand people for high income (HIC), middle income (MIC), and low income (LIC) countries, 1988–2004 (*source*: World Development Indicators).

Table 9.1. *Criticism of the Global Billboard Society (after Hamelink, 2004: 75)*

Global Billboard Society	International Human Rights Imperative
Maximising corporate profit	Optimising public welfare
Treating people as consumers	Treating people as citizens
Private appropriation of public space	Safeguarding public space
Priority for trade-law principles	Priority for human rights
Culture as commodity	Culture as public good
Ignoring inequality in market transactions	Aspiring equality in human interactions

this privatisation tendency is less strongly witnessed, although public states are also here losing media ground. Or, in Stevenson's (2001: 63) words, "The rapidly emerging information highways, the multiplication of television channels, the increasing power of communication conglomerates and the development of media technology are all driven by the instrumental logics of science and profit." Hamelink (2004) has coined this Global Billboard Society, and identified its characteristics vis-à-vis his preferred International Human Rights Imperative (see Table 9.1). In parts of Asia (such as the Middle East, China and Vietnam; cf. Chapter 10), the national media are still strongly in the hands of the state, limiting not only the commercialisation and profit-drivenness of information and news, but also the diversity and democracy of news and information.

The third development that changes the role of media and information gateways is the emergence of the Internet and cyberspace. Although many scholars are rather critical or even negative on the former developments of global spreading and concentration of commercial media conglomerates, most are much more positive of the developments of the new electronic communications (cf. Paehlke, 2003; Tumber, 2001). The new media that run via cyberspace and the Internet often are seen as contradictory to and counterbalancing some of the problematic tendencies that characterize the 'old' media of television and newspapers. Whereas the old media often are interpreted to be linked to, and structured and determined by, transnational corporations and – to a lesser extent today – state bureaucracies, the Internet is believed to have grown more organically in relation to science and civil society. Although we have seen in former chapters that this is indeed to some extent correct, the role of multinationals as Microsoft, Google and AOL balances a too black-and-white division in the mediascape. With respect to cyberspace and the Internet, the struggle over power and control has only just started, and it remains to be seen to what extent the current democracy and relative lack of monopolisation of the Internet will endure in the future. Some see similar dangers for Internet as what they identify for the current conventional media: state-control in nondemocratic societies and capital control in the OECD market economies (cf. Paehlke, 2003; Downing, 2005). Goldsmith and Wu (2006) even conclude that within democratic societies there is increasing state control over the Internet, making a borderless Internet an illusion. Others are less pessimistic and witness also innovative developments, whereas the new media challenge the existing conventional media in their monopolies: Indymedia (cf. Chapter 8), the Korean Internet newspaper OhmyNews[5] or the diffuse fluid of webloggers.

The media as battleground

Overall, over the past two decades, the media "have become more powerful than ever technologically, financially and politically as their global reach allows them to escape from strict political controls" (Blumer and Gurevitch, 1995). Increasingly, the media have managed to move beyond political control, not only through privatisation and monopolisation but also through the globalisation of media and informational flows, making nation-state institutions ill-equipped to govern transnational flows. In that sense, the 1970s ideas of the NWICO on state control of incoming and outgoing information flows would sound naive

[5] This Internet newspaper, established in 2000, is produced by thousands of citizen-reporters, who write articles that are corrected, verified and edited by the publishing house, which also has the right to refuse articles (http://english.ohmynews.com). Similar experiments in other countries have followed this example (e.g., http://www.janjan.jp in Japan; http://www.mymissourian.com in the United States), often with less success in challenging the conventional media because of the lack of the specific Korean historical backgrounds of information self-censorship in the 1980s.

in the twenty-first century, as even countries such as China and Vietnam increasingly have difficulties to remain in 'informational control' (cf. Chapter 10).

But the idea of an all-powerful media beyond any control and governance, a Fourth Estate, is too simple and monolithic. Paehlke's (2003: 93–95) idea of global media corporations that increasingly develop the capacity to influence and even control how the environment moves on the public and political agendas, and thus subvert the issue attention cycle, is a too-massive monopoly-capitalism analysis. In criticising the idea of both increasing instrumental rationality of science and profit in the media, and an overall loss of hermeneutics, Stevenson (2001: 64) emphasises the other side of the media coin: the liberation that the media has brought for the oppressed. The future may very well combine the increasing power of multinational media conglomerates with the new possibilities for interrupting and subverting the dominant discourses of the powerful by citizens, communities and new social movements. There is no one evolutionary and necessary development path of the media towards just further monopolisation, commercialisation and homogenisation. The media is socially constructed, and its development path and outlook depends on the various forces struggling for media control and hegemony, among which are the media conglomerates themselves. Stevenson (2001) identifies two potential strategies that can and do challenge the dominant agendas and power of global commercial media. The first is the development of a full-fledged 'global' public sphere underpinned by widely recognised rights for diversity and access to communication, as, for instance, put on the agenda by Hamelink's (1995) proposal for a worldwide People's Communication Charter. The second consists of constant interruptions and infringements on the commercial global media by social movements, nation-states and international public organizations, in order to limit the commercial and monopolistic hegemony of a few.

As society comes increasingly to its citizens via screens, and politicians, parties and movements have to reach society via acting in and through the media, the media becomes important as a battleground for power struggles and politics, rather than only a power on its own. One can only fully understand that if the idea of a homogeneous media is broken down: media are diverse, complex and multidimensional (in news production, in the kind of media through which news/information is transferred, in geographical reach, etc.). And the media do not produce homogenising messages for a global village, but produce differentiated products adapted to place-bound characteristics for different audiences. As such, media politics includes both oppression and resistance, as Castells (1996/1997) illustrated so colourfully in his opus magnus. So it should not surprise us, as Stevenson (2001) shows in reviewing the various interpretations and judgements of the current developments in and of the media, that one can find media pessimists and optimists, criticism and praise, of all sorts from all kind of political directions.

With the media as a new battleground, rather than just a Fourth Estate, the crucial position journalists had is changing as well. How should we define the role of journalism in the Information Age? Many agree that the traditional role of journalists and news editors as gate-keepers to the world is no longer adequate and preferred in an age marked by both global commercial media monopolies and the diffusion (or even democratisation) of information production and consumption via the World Wide Web.[6] Others even doubt whether journalists and newsworkers, even if they would prefer to continue in those roles, would be able to play such a role in current times. In analysing the information and communication flows around the Starr report (in 1998)[7] and the 9/11 event (in 2001) in the United States, McPhail (2006: 301) concludes:

The important communication point to be made in all of this is that no longer news editors, pundits, politicians, the US president, or others are able to act as gatekeepers to restrict, alter, spin, or limit the information in the report. Rather, millions of average people around the world now have access to the full, unedited, government reports at the same time they are presented to the national legislatures.

If these conventional roles of journalists as newsbringers and gatekeepers are no longer valid, what are then new roles for (environmental)

---

[6] In stressing the consequences of commercialisation in the media, Paehlke (2003: 80–81) quotes from a survey among journalists in the United States: 40 percent of the journalists practice self-censorship by not covering stories that might offend advertisers or the owners of their news organizations; 52 percent avoided too-complex stories; 77 percent turned away from stories that were important but dull. These figures should not be too surprising for environmentalists having experience with the difficulties of getting complex and unwelcome environmental messages across mass media.

[7] The report of the independent counsel Kenneth Starr on the Monica Lewinsky affair, which was followed by an impeachment procedure on U.S. president Bill Clinton.

journalists? Glasser and Craft (1998) refer to the emergence of public journalism in the late 1980s in the United States, where journalists and the media have a role to play in revitalising the quality of the public discourse and with that of an active democracy. Bardoel (1996) sees a new role for journalists as information brokers, providing functional and specialized information to interested citizen-consumers. Others draw even more radical consequences from the fact that there are no longer any gates to keep. The Internet contains the multiplication of information sources as well as information producers; thus, the emergence of People's Journalism. If anything, journalists have only a (public) service role to play, in bringing together the various sources with the various consumers (Yelvington, 1999), as facilitators of public discussions, as connectors between supply and demand (Aufderheide, 1998), as the ones opening up critical questions for discussion (Stevenson, 2001). With the evaporation of these last authorities and landmarks in the information and media arena, news and information can indeed best be understood as fluid flows of information without any predetermined routes, node and networks and without any clear centre of governance or control.

5. Environmental politics and the media

How do these developments and transformations in the media work out for environmental governance? What is the relevance of all this for understanding informational governance of the environment?

In assessing and explaining what they call the 'culturist turn' in recent political campaigns and collective actions of environmentalists and other movements (such as those of the anti- or other globalization movement; cf. Chapter 8), Scott and Street (2001: 45ff) turn away from a cultural explanation. According to them, the fact that environmentalists and other social movements turn away from political institutions, political structures and political actors as the focus of their activities, claims and strategies and move towards the media, the form and visualisation of their messages, the clashes of different cultures and an emphasis on blaming and shaming the economic and political powers against universal norms and values should not be explained in cultural terms but in 'classic' terms. New resources, new rules of the 'political' game and new opportunity structures lead new social movements to redirect their focus, strategy and collective actions. In an age of information, the conventional media (press and television) and the new media have become crucial resources and battlegrounds, and this also redefines the strategies and activities of new social movements. But it does not turn new social movements into cultural movements, and one should not explain these developments in terms of a new culture of protest and engagement.

In analysing the media resources and battlegrounds, the old media differ strongly form the new media. The conventional media have always been looked on by the environmental – and other social – movements with suspicion. These media were and still are seen as a homogeneous, hostile force, which rather misrepresents and distorts the messages of the movement, while leaning particularly strongly on governments and corporations in news reporting. Conventional media tend to induce professional management of the media, professional and private ownership of the media, and concentration of the media in a few hands that run to a major extent parallel with capital. But, at the same time, and increasingly from the late 1970s onwards, the environmental movement became fundamentally dependent on the mainstream mass media, as these media turned out to be a strategic resource. Without this mass media, the (Western) environmental movement was probably not able to develop so rapidly and could not have such an influence in the national and global political game. The current environmental battles for media attention and information dissemination to the public is characterised by professionalisation, a protest business, a growing similarity between commercial campaigns and those of environmentalists, and the prevalence of form over substance. The classic example of this is Greenpeace's media campaign concerning the Brent Spar (cf. Bennie, 1998; de Jong, 2005; for more general press coverage of Greenpeace, see Hansen, 1993a). A more recent, but even more powerful, example is the 2006/2007 media campaign around and by Al Gore's Oscar-winning movie/documentary *An Inconvenient Truth*, which caused major accelerations in the climate change agenda-building, policies and measures worldwide.[8] Contradictory to the Brent Spar case

---

[8] *An Inconvenient Truth* followed to a major extent the logic of conventional media conglomerates, with large funding and a highly professionalised media campaign. But, at the same time, it was more a documentary than a movie, and in that sense it was very different from the 2004 Twentieth Century Fox movie *The Day after Tomorrow*, which also addressed climate change issues but did not have this tremendous and global impact.

and Al Gore's documentary, however, in most cases environmental NGOs do not fare well in dominating news production.[9] Conventional media have a clear preference for governmental, political and corporate institutions (Manning, 2001; Cottle, 2000; de Jong et al., 2005; Hannigan, 2006). Complex, slowly developing environmental disasters brought to the media via repetitive stories and campaigns of environmental NGOs, such as climate change, have generally little attraction for sound-bite media, even with the scientific and governmental communities supporting (and often paying for) such news coverage.

The Internet opens a quite distinct set of media opportunities and battles on the environment. Essential is that the Internet is multicentred and not (yet) monopolised, and information production and dissemination are affordable also for nonwealthy actors. Moreover, the Internet remains difficult to control and access, and production and dissemination of information take place largely outside the control of global media corporations and governments. Although media sceptics warn us that commercialisation also may get hold of the Internet, they seem surprised that this has still not happened: "It seems at times almost too good to be true in terms of the extent to which there may now be wide access to a medium not utterly dominated in terms of content by the smooth voice of infotainment and large-scale capital" (Paehlke, 2003: 103). Although professionalisation still plays a role in the new media (e.g., in terms of being able to built and manage Internet sites), this is less dominant and monopolistic as with the old media (Scott and Street, 2001: 46). Equally, capital needs for news production and dissemination with the new media are a fraction of what they are with the conventional mass media. It is clear that the preconditions of the new Internet battleground for environmental advocates are much more advantageous than those of the conventional media. And, indeed, the Internet traditionally has been dominated more by social movements of different kinds than by state authorities and TNCs (with the exception of the Internet-related companies themselves, of course). Only recently, corporations and governments are discovering the potentials and powers of Internet (cf. Chapters 6 and 7).

---

[9] Also within the environmental movement and among activist groups there are clear differences in media access and influence, although not always along the lines of member size, economic resources and political power. In that sense, media have their own rationalities, which cannot be reduced to mere economic or political ones.

The emerging importance of the Internet as media gives also rise to new activities and actors, such as Internet surveillance firms that monitor what is being said on the Internet by competitors, critics, supporters and rumours. Such surveillance firms make up reports for their clients, which can be followed by adequate action strategies, depending on the kind of presence. Others rely more on active linking to operate strategically in the new media, which has been interpreted lately as a wider shift from a so-called hit-economy to a link-economy on the Internet. If one searched for Greenpeace via the search engine Alta Vista, it would result immediately in advertisements on the contribution of Shell to greening the world on your screen (Rogers, 2002). But linking also has everything to do with piggy-backing on the reputation of other sites and organizations, or breaking down such reputations. In that sense, reputational capital has not really been changed by the Internet.

Still, a number of dangers with respect to the new media remain: privacy intrusion, large-scale Internet control (especially after 9/11, and in some authoritarian regimes), growing inequality in access, and a development towards 'television and entertainment with advertisements' as the axial principle around which the Internet is organised rather than news, information and communication. In that sense, the unique characters of the Internet are not pregiven or intrinsically connected to it, but need to be actively protected and sustained. The global media concerns that dominate the old media actively aim to integrate the old and the new media; and concentrations such as those of Google, Microsoft, Skype and other Internet companies are equally challenging the democratic properties of news and information production and dissemination, as a number of states do. But that should not be surprising, with the growing importance, influence and agenda-setting of the Internet in economic, political and cultural life.

Of crucial importance of the relevance of the new media in environmental politics are two factors. First, which media are citizen-consumers, and to a lesser extent manufacturers and governmental agencies, consulting in their search for entertainment, information, news consumption, public opinion formation and so on? To what extent will the new media indeed replace the old media in some or all of these functions? Second, how important will be the spill-over of the new to the old media, and of the old media to the new media? Is information circulating through the Internet a new rich source that has a reasonable chance of spill-over to the conventional media, strongly enhancing the powers of transparency and accountability? Or will the old media, with its inequalities in information and news production, definition and transmittance, become a dominant factor in shaping and structuring the new media?

6. Conclusion: media as governance, governance of the media

In the Information Age, the media has become a key player in environmental governance, as much as it has become a key factor in many other fields. In searching for informational governance in environmental protection we cannot but acknowledge the fact that environmental governance has become mediatised. But, at the same time, the media can no longer be simply seen as a limited set of multinational players that have an all-encompassing power over key governing processes. Of course, power in the mediascape is unequally distributed, with too much in the hands of a few multinational media monopolies or oligopolies. But this is not the essence of the role of the media in contemporary environmental governance. Rather than interpreting media companies as key actors that govern environmental (non)protection, the media should be seen as a key battlefield on which environmental politics, conflicts and priority setting are increasingly being settled. In addition, the conventional media of television, radio and the written press are more and more challenged by – and integrated with – the new media, which significantly changes the mediascape. The power balances around these new media are not settled yet, and states, old media conglomerates and a wide set of new actors appear on the production, distribution and consumption stage of these new media, among them environmental advocates. In these new settings, and especially with respect to the new media, actors become integrated entities, switching between producers, distributors and consumers/audience of news and information. One could indeed label that democratisation, but of a specific kind, as many are still excluded. A more critical analysis would instead point at new lines of inequality, now strongly defined by new media rationalities.

So, the key question increasingly becomes: how is the media governed? Although throughout this book we have become aware of how important processes of transparency, verification, disclosure, accessibility and accountability are in the new forms of environmental governance, and that the – old and new – media play a key role in this, we increasingly have to look with these very same criteria at the media. How are the media doing in terms of transparency, verification, accountability and disclosure when they are involved in informational governance processes (also on the environment)? The new media were initially celebrated for their contribution to information diversity, access, transparency and disclosure, especially vis-à-vis the old media. Whether that holds – especially under conditions of a further integration of old and new media – is one of the key critical questions for future media research, also with respect to environmental governance.

10 | *Information-poor environments: Asian tigers*

1. China and Vietnam as information peripheries

Until this chapter, most of our analyses on the consequences of the information revolution for environmental governance have been focused on the more developed parts of the world, or on the global networks and flows that connect and integrate these developed parts of the world with specific nodes, places and practices in the lesser-developed countries and regions. It goes without saying, however, that information flows, informational processes and information access are not similar in all parts of the world and in all countries (cf. Zook, 2001; Gunaratne, 2002), resulting in geographical variations in the importance and relevance of informational governance on environment. The limited number of studies and analyses on how the information revolution and the Information Society affect environmental protection and governance have focused on the more developed parts of the world (see, for instance, the studies on the Information Society and sustainability by the EU). These are all information-rich environments, in which information generation, processing, access and use are relatively abundant; and – consequently – where informational dynamics can indeed play a significant role in environmental governance. But as this widespread availability of (environmental) information is not found everywhere and significant parts and spaces of the globe witness severe information shortages, what does informational governance look like in information-poor environments? Is informational governance at all a relevant category in such environments? In order to start gaining an insight in these questions we will look at two developing countries: China and Vietnam. But we start in this section to put our case study entities in perspective.

In current times, information is often connected to ICT and the Internet. It has been especially the notion of digital divide that has coined the dissimilarities in ICT and Internet access around the globe from the early 1990s onwards. And still the figures are telling: large parts of developing countries, and especially those in sub-Saharan Africa, are largely disconnected from the World Wide Web. The digital divide points our attention to only one specific dimension of informational governance (digitalised information related to ICT and Internet). Although related to the Internet and ICT, informational governance on the environment should not be limited to digitalised information flows. It also has to do with nondigitalised monitoring and information collection programs, with transparency and disclosure of information, with monopolising information, with informational controversies. Exactly for that reason, Chapter 1 identified four dimensions of information-poor-environments: economic, political, institutional and cultural. In developing countries, we often will find mixtures of economic, political, cultural and/or institutional causes of information-poor environments.

More recent analyses of the informational highway and the information hubs in the global economy have fine-tuned the rough geographical division in the information-rich OECD and the information-poor developing countries. The global informational networks and flows connect and integrate developed parts of the world with specific key nodes, hubs, places and practices in the lesser developed countries and regions. Although we can still claim that sub-Saharan Africa and developing countries in Asia are poorly connected to the World Wide Web, specific localities and specific practices (productions linked to transnational commodity chains; members of transnational NGO networks) in these regions manage to connect well to the informational highway or even become a hub (e.g., metropolises). And specific locations, practices and groups in more developed countries are falling short of information flows, as we have noted in the previous chapters.

So, although information-poor environments cannot be equalised with developing countries, it is most likely that (regions, practices and groups in) such countries will face informational deficits and show characteristics of informational peripheries. Although some of the (four) causes behind information-poor environments also can be found in the geographies and spaces that are closely linked to the global informational networks and flows, the consequences will be less severe and extreme. In the empirically informed analysis on informational

Table 10.1. *Internet users and telephone lines per one thousand inhabitants and FDI flows in US$ per capita*

	Internet users per 1,000 inh. (2005)	Fixed and mobile telephone lines per 1,000 inh. (2004)	FDI per capita (in US$; 2002–2004 average)	
			Inflow	Outflow
North America	681	1223	263	591
Europe	359	1430	445	473
Asia	99	431	27	10
China	85	499	38	1.3
Vietnam	70	131	18	?
Latin America	143	499	98	20
Africa	25	84	19	5

Source: ITU; World development Indicators; http://www.geohive.com/.

peripheries in this chapter, I will restrict myself to two countries – China and Vietnam – to explore how informational peripheries are affected by the new modes of informational environmental governance. But we do that with the knowledge that it is only one empirical category (developing countries) of informational peripheries; and more: within that category the case study examples are not representative.

The choice for these two countries is, of course, arbitrary. China and Vietnam are far from representative for the set of non-OECD countries that are not fully located in the prime nodes of informational flows and ICT developments. Arguably, they differ strongly from especially a significant number of sub-Saharan African countries, which seem to be much more peripheral in the global network society and economy, for instance when we consider the number of Internet connections or the number of personal computers (but less so for foreign investments; see Table 10.1 and Figure 10.1).[1] In addition, both countries deviate from a considerable number of non-OECD countries in that they come from – and still have some of the characteristics of – a centrally planned economy with specific – although changing – relations

---

[1] Consequently, environmental monitoring and information systems are much less developed in these African countries, compared to China and Vietnam. See, for instance, Prévost and Gilrith (1999) and Gavin and Gyamfi-Aidoo (2001) on African environmental information systems.

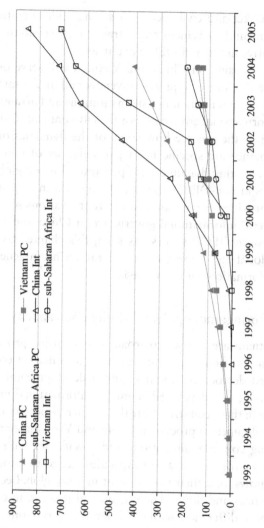

Figure 10.1. Personal computers (PC) and Internet users (Int) per ten thousand inhabitants in China, Vietnam and sub-Saharan Africa, 1993–2005 (*source*: World Bank Indicators).

among state, market and civil society. A strong undemocratic state, a weak environmental movement and a transitional economy are far from representative in the non-OECD countries.

Nevertheless, focusing on China and Vietnam will serve our purpose if we leave the idea of representativeness and at the same time go beyond a focus on only nation-state institutions and arrangements in analysing informational governance of the environment. China and Vietnam enable us then to see how some of the dynamics of informational governance are specific for a particular set of non-OECD countries, whereas other dynamics also penetrate through global networks and flows in non-OECD regions. Or to put it in Appendurai's vocabulary: it provides us with an idea where the various scapes that allow and 'govern' informational governance in China and Vietnam differ from Western OECD scapes. As such, this chapter gives us a taste of how global and homogeneous (or rather: heterogeneous) at this moment informational governance is.

2. State monitoring: monopoly, reliability and capacity

China and Vietnam are at the crossroad of two major processes: a transitional process (from a centrally planned to a market economy, with still limited democratic liberties) and a developmental process (from a developing to a developed country). Informational processes and governance in these countries bear the characteristics of both processes. The developmental process in China and Vietnam shares with other developing countries limitations and specificities with respect to informational infrastructures and capacities, access to information networks and flows, and their involvement in more global economic processes, networks and flows. The transitional process, in contrast, brings a number of differences with other developing economies. First, the close alliance of state-owned components of the economy with parts of the state bureaucracy still make stringent environmental monitoring, control and enforcement a complicated matter. Often environmental authorities could not and cannot operate independently in enforcing environmental laws and regulations; they need to negotiate and discuss interventions with the more powerful parts of the bureaucracy. The growing separation of state and market to some extent changes this (cf. Lo and Tang, 2006). Second, reliable information has always been one of the major problems of centrally planned economies. With numerous incentives at all levels to distort information, limited transparency in information collection and processing, hardly any independent control on the reliability of information, and very limited independent information sources outside the government, adequate information was never guaranteed (Lieberthal, 1995; Ma and Ortolano, 2000). This strongly influenced environmental governance. With the transition process on its way, this is changing somewhat (although to different degrees in the two countries), but the transitional stage still affects the outlook and workings of informational governance. Third, in the centrally planned economies of Vietnam and China, formal organizational structures, arrangements and legal regimes could constantly be bypassed by party politics, informal networks and unclear decision-making structures. This jeopardized consistency, transparency and reliability in policy making and implementation, also in the field of environment. This is only changing slowly in the transitional economies of today, and party politics continue to work along parallel lines and arrangements of decision making. This also enhances the risks (and practices) of corruption and bribing in (environmental) policy making, exactly because transparency and consistency are lacking. Finally, the transition process has its consequences in the area of civil society and democracy. In Chinese and Vietnamese environmental governance, civil societies play only a marginal role. In both countries, there are hardly any nationally organised environmental NGOs that can put pressure on the policy-making and the economic systems to ecologise. Although in China we have witnessed the first sprouts of national and local environmental activism and the media seem to be given more freedom in reporting on environmental misbehaviour, in Vietnam no such developments have been observed at the time of writing (Mol and Carter, 2006; see later in this chapter).

Monitoring in Vietnam

Since 1995, Vietnam has a national environmental quality monitoring system, in which the country is divided into three regions: North, Central and South. One environmental research institute is responsible for environmental quality monitoring in each region: CEETIA (HaNoi Civil Engineering University) for northern Vietnam, the Environmental

Protection Centre (EPC/VITTEP) for central Vietnam, and CEFINEA (HoChiMinh City University of Technology) for southern Vietnam. Environmental quality monitoring of water and air takes place regularly, focusing on a limited number of parameters.[2] The National Environmental Protection Agency NEPA, within the Ministry of Natural Resources and Environment MONRE, collects these environmental quality data and reports on them, among others via the recently started annual State of the Environment Reports.[3] There is no national monitoring programme on emissions of point sources such as major industries or power plants, nor on diffuse sources such as agriculture or motorised traffic. In addition to this national monitoring system a few large cities (especially HoChiMinh City and HaNoi) have municipal air and water quality monitoring systems, financed via local funding. We will focus on HoChiMinh City as an example, being aware that this is arguably the best practice in local environmental monitoring in Vietnam.

HoChiMinh City has eight surface water quality monitoring sites and nine air quality monitoring sites spread around town. The latter are set up in 2002 with Official Development Assistance from Norway and Denmark. The local environmental monitoring system in HaNoi is less well developed and meets several technological, organisational and managerial constraints in implementation (Pham Minh Hai, 2005). Reports on processed monitoring data are published with some delay and most of them are only available on request from the monitoring

---

[2] Environmental monitoring for NEPA by these institutions includes surface water quality, air quality and solid waste. For each of the three regions, there are ten surface water monitoring locations (some twenty parameters are included), six locations for monitoring air quality, and one for solid waste. NEPA budgets for environmental monitoring have increased sharply in the past few years, especially because of the introduction of automatic monitoring systems, and not so much following expansion of the number of monitoring sites. There are a number of other ministries that have monitoring on environmental issues (such as those on biodiversity, on pesticides, energy), but there is little data exchange and coordination between the ministries responsible for the various monitoring programs (Pham Minh Hai, 2005).

[3] A four-hundred-page long SOER is annually published for use within NEPA, a shorter ninety-page version is sent to the National Assembly, and a summary is sent to other line ministries (Faucher, 2006). In 2005, for the first time, a concise version of the State of the Environment report was made publicly available, which is part of new legal requirements as of 1 July 2006.

agencies or from the responsible authorities. Although limited in number of parameters and frequency of sampling, interested parties can have access to environmental data.[4] Often, environmental data are treated as an additional source of income for officials, and off-the-record payments can result in surprising data sets. There is use of conventional media, but hardly any use of the new media, in making access more easy and widespread. The exception is the Air Quality Index at the Web site[5] of the HCMC Environmental Protection Agency (HEPA), also presented at a large digital bulletin board at the central market in HCMC, alternated by various commercials. Governmental Web sites are fragmented in environmental data disclosure, often presenting one-time environmental quality data on a limited number of parameters.

This all looks still impressive compared to industrial (and other) emission monitoring. There is no general system of systematic monitoring of industrial and other point source emissions by the local governments in HoChiMinh City and HaNoi (and certainly not outside the major urban-industrial nodes). Since early 2004 HEPA is the responsible agency under the HCMC Department of Natural Resources and Environment DONRE for emission monitoring. Both before HEPA's establishment and at the moment there is no systematic collection of data of industrial emissions (e.g. Tran Thi My Dieu et al., 2003). Recently, HEPA started a first attempt to get emission data from the 100 major industrial pollution sources, but only 40 industries reacted by sending HEPA (some of) their emission data. Overall, reactions from foreign investors and joint ventures (which are better linked to the informational highway) were much better than from major domestic companies. From this a GIS-based data system is under construction, which is only available and accessible for governmental

---

[4] The HoChiMinh City Environmental Protection Agency (HEPA) did discuss recently with various local media whether the latter could not publish the so-called Air Quality Index regularly in their newspapers. These newspapers refused to do so, using the argument that the data were not reliable enough (interview HEPA official, October 2005).

[5] HEPA presents an air quality index on their Web site (http://www.hepa.gov.vn/AirOnline), be it with a time lag of some two weeks because of the need to process raw data and make data more reliable. There is also opposition from higher echelons, both locally and nationally, against the (further) environmental disclosure policy of HEPA, limiting HEPA's possibilities to expand in this direction (interview HEPA official, November 2005).

agencies and research institutes upon request, but not for other interested parties. There is no available information on industrial emissions other than these (incomplete) collected and monopolized data of the government. HEPA is also unable to press the other 60 major industrial polluters to send their emission data, as there is no strong legal basis under this information collection.[6] In addition, any pressure on industries to provide data has to go via the Environmental Management Agency of DONRE, who is more powerful as this agency has to issue environmental licenses. But also this agency has limited possibilities to push industries to release emission data, as it would soon lead to interventions from the People's Committee of HCMC to reduce pressure in order to enable economic development. Obviously, lines are very short between the major industrial polluters and the People's Committee in HCMC. The situation is in principle not different in HaNoi (Pham Minh Hai, 2005)

One of the consequences from this lack of (availability of) emission data is the poor rate on environmental fee collection. At the moment HCMC only knows a waste water fee, based on the water pollution load.[7] The fee is established based on information provided by the company, following the return of a filled-in form to HEPA (especially to its Division of Administration, Finance and Environmental Fee Collection). In principle, HEPA checks the data provided by the company by comparing these data with general information on emission factors

---

[6] The 1994 Law on Environmental Protection provides the overall framework for environmental policy and management in Vietnam and has been detailed since then with a significant number of decrees. But still it does not provide a sufficient legal basis for setting up an adequate, extensive and detailed system of environmental monitoring and information gathering. There is an obligation for major companies (both private and state-owned; both domestic and foreign ones) to send annually an environmental report to HEPA, but few do so and there is no strong follow-up on noncompliance. O'Rourke (2004) noticed the same for industries in HaNoi.

[7] Of this fee, 15 percent goes to sampling activities, 5 percent of the fee goes to the administration agencies (of that 1 percent to HEPA and 4 percent to the district Environmental Boards) and 80 percent of the fee goes to an environmental fund. Of this 80 percent, half goes to the national environmental fund and half to the local environmental fund. Unofficially, DONRE of HCMC has recently (2005) agreed with MONRE that HCMC can use all of the 80 percent, without any part flowing back to the national level. However, in HCMC, these contributions for the local environmental fund were in 2005 mainly used to pay back loans for environmental infrastructure to the ADB, leaving no money for other industrial environmental improvements.

of production processes and available data on production levels and outputs. But in practice the system shows many shortcomings. Only a small fraction of the forms sent out is returned. Because of limitations in manpower (with only 5 persons working in the fee collection division of HEPA for the whole city, and some part-time work of staff in the district environmental boards) the filled-in forms are hardly corrected and the firms that have not sent in their forms are hardly reminded or prosecuted. The end result is a very low level of fee collection at the city level. Outside the main urban-industrial nodes, industrial emission monitoring is even less systematic and data are hardly available even for governmental agencies and research institutes. But remarkably the fee collection is better, due to lower levels of industrialization and thus lower work loads of the environmental authorities.

The alternative for low levels of actual monitoring of emissions would be model calculations based on more general parameters. However, the availability of environmental data based on modelling and calculations is poor, due to the lack of information and knowledge on data that need to feed models on emission calculations, such as production process characteristics, production volumes, emission parameters, natural resource use, energy and water use, etc. Most environmental data follow still from place-based sampling and monitoring, which happens incidentally and not in a structural and systematic way over longer time periods.

Monitoring in China

China differs in some respects from Vietnam in environmental monitoring, data collection and availability and transparency of environmental data. Systematic data collection programmes on environmental quality have been extensive and national environmental information is more widely available, especially through the Internet. If we compare the Web sites of the Chinese State Environmental Protection Agency (SEPA) with the Vietnamese NEPA,[8] or the statistical annual environmental

---

[8] See http://www.sepa.gov.cn for China (and http://www.zhb.gov.cn/english/SOE for the English versions of the annual State of the Environment reports) and http://www.nea.gov.vn for Vietnam (and http://www.nea.gov.vn/english/state.htm for the English version of the State of the Environment report 2001 and recent monitoring data). From 2005 onwards, the China Environmental Yearbook also has been published in English.

reports or yearbooks of the two, major differences can be witnessed in the amount of environmental data, the details of these data, the time lines in all kind of monitoring programmes and the number of different indicators that are monitored. Part of these differences is related to the mere size of China, part also to the higher economic levels and environmental state capacities in some of the large cities and eastern provinces. But the wider public availability of environmental data in China is of recent origin and was strongly triggered by the 2003 SARS (Severe Acute Respiratory Syndrome) experience in which governmental agencies were held accountable for the secrecy of health data and dangers. Increasingly, environmental data are no longer seen as sensitive and are opened for the wider public, also through the old and new media.

Nationally, the Environmental Monitoring Centre (EMC), based in Beijing, plays an important role in data collection and data processing, whereas the SEPA is responsible for environmental data publication and disclosure. With around twenty-three hundred environmental monitoring stations in more than 350 cities, China has an extensive monitoring network.[9] The EMC is responsible for quality control and certification of the monitoring stations and is financed from the state budget via SEPA (some US$8.5 million in 2005). The EMC also partly finances local monitoring institutions (for which it receives an additional US$8.5 million annually), but only for those tasks that are related to the national monitoring system.[10] There is very little exchange of

---

[9] These are all governmental monitoring stations, belonging to different governmental levels. In addition, there are more than twenty-six hundred monitoring stations belonging to various industries. On a regular basis, data are collected on four themes: city air quality, surface water (rivers and lakes), drinking water and noise levels. Local monitoring data is collected at the provincial level and then send to the National Environmental Monitoring Center. After data processing EMC prepares annually around thirty different environmental data reports, of which the State of the Environment and the China Environmental Yearbook are widely available. All other reports are not disclosed, but usually available for SEPA, the State Council and a few other governmental institutes (interview deputy director EMC, November 2005). See also the APEC Virtual Centre for more information on the China environmental monitoring stations (http://apec-vc.cestt.org.cn).

[10] For instance, Beijing has twenty-two monitoring stations, of which only seven are included in – and thus financed by – the national monitoring network. Waste monitoring, for instance, is not included on a regular basis in national environmental monitoring.

environmental data between the EMC and SEPA, on the one hand, and other ministries (such as those on water, on forestry, on energy), on the other. Environmental data are also treated as a commodity, quite similar as in Vietnam.

If we compare, for instance, environmental monitoring in Guangzhou,[11] a rich metropolitan city of seven million inhabitants in southeast China, with that of HoChiMinh City, similarities and distinctions can be noted. Irregular water quality monitoring in Guangzhou started in the 1970s, but more systematic monitoring started from the mid-1980s onwards. At the moment, surface water quality monitoring takes place six times a year on fifteen points regarding twenty parameters, quite comparable to HCMC water quality monitoring. Monitoring of emissions at industrial sources is, however, quite distinct. In 2002, emission data of twelve hundred industries were collected in Guangzhou,[12] as well as discharges from domestic waste water. Monitoring is related to the control and enforcement of discharge permits, which were first implemented in Guangzhou in the late 1980s, as well as to collecting pollution charges (which can be reinvested in environmental technologies of the company up to 80 percent).[13] Environmental quality data, the annual environmental quality report, as well as the examination results of the Urban Environment Comprehensive

[11] This example draws on research carried out by Jinyang Zhang (2005) at the Environmental Policy department at Wageningen University, the Netherlands.

[12] This included parameters such as ammonia, COD, heavy metals, oil products, cyanide and some POPs. Nationally, over the years 1996–2002 only 55 to 60 percent of the companies included in the National Statistical Bureau are included in the environmental statistics and monitoring. These companies under environmental statistics and monitoring represent also around 60 percent of the total Industrial Output Value of China (Shi and Zhang, 2006). Emission monitoring is not done systematically by governmental agencies but follows the legal requirement of companies to report their discharges. The EMC does not collect these data systematically (interview deputy director EMC, November 2005).

[13] In 2002, around RMB670 million was collected in Guangdong province from pollution charges. One-third came from Guangzhou and of that around half was from water pollution charges. Although 80 percent of these charges have to be invested in environmental projects and improvements in the companies, 20 percent goes to the budgets of the provincial and local Environmental Protection Bureaus. By 2005, the allocation of money from environmental charges was changing countrywide, first in the richest provinces. See, for more information on how that worked out in Guangdong province and its counties, Lo and Tang (2006).

Improvement Quantitative Examination System[14] have to be made available to the public, usually via state controlled media and the Guangdong EPB Web site. Every major Chinese city also publishes an Air Quality Index. This public disclosure is not the case for the emission data of the individual companies.

Unlike Vietnam, China is using these monitoring systems for their rapidly developing system of environmental auditing, with the China National Audit Office in Beijing covering national policies and projects and with provincial (and sometimes even municipal) audit offices more locally.[15] Until recently, environmental auditing was limited to financial auditing, but increasingly performance auditing is in development. These auditing offices are, rather, independent from direct policy-making and implementing agencies and report directly to the State Council (and its equivalents). Increasingly, they have a major influence in improving state policy making, avoiding illegal spending of funds, reporting governmental misbehaviour and even controlling environmental implementation.[16] Mostly, environmental auditing reports are not made public in China, and thus they have played a limited role in informational governance until now. China's growing involvement and active participation in INTOSAI, the international network of national Courts of Audit, may also, in the near future, result in increasing attention to transparency and public disclosure, according to representatives of the China National Audit Office.

---

[14] UECIQES is an examination system to compare how well municipalities and its leaders are doing in combating environmental pollution. In 1999, Guangzhou ranked twenty-eighth among forty-six major cities in China and tenth among twenty-one cities in Guangdong province (see, for a further explanation of this system, Rock, 2002). This comes very close to a national equivalent of the global Environmental Performance Index and data-driven regulation arguments as developed by Esty and colleagues (Esty, 2001a; Esty and Rushing, 2006; cf. Chapter 6).

[15] Auditing, and also environmental auditing, at the provincial level can be quite large. The Autonomous Region of Inner Mongolia (twenty-two million inhabitants) has twenty-three hundred auditors (2005 data); 115 of them work at the division for Audit of Agriculture, Natural Resources and Environmental Protection. Environmental auditing had an impact in improving the wastewater discharge fee system, combating bribing and improving the program of conversing agricultural slopeland to grass and forests (interview, official at the Audit Office of Inner Mongolia Autonomous Region, November 2005).

[16] Interview, deputy director Audit Research Institute, China National Audit Office, November 2005; interview, professor at the Chinese Academy of Sciences, Beijing, November 2005.

China's monitoring capacity is also well illustrated in the recent establishment of the China Biodiversity Information System, a large information system run by various research institutes of the China Academy of Sciences. This system is strongly digitalized, with more than one million records, and forms the basis for protecting more than two thousand protected areas (15 percent of land cover) and the running of the China Species Red List.[17] The international Convention on Biodiversity has strongly triggered this monitoring system.

If we do not pay attention to the national level and the more wealthy eastern provinces and cities and move instead to poorer regions and local levels, a much less favourable picture emerges of environmental monitoring and information in China:

- scarce environmental monitoring as a significant part of environmental monitoring needs to be funded by the local governments, who have limited budgets and different priorities;
- distortion in information processing;[18]
- secrecy and commoditisation of environmental data for large segments of society, also, for instance, for scientific institutes;[19]
- absence of a right-to-know code, legislation or practice, both at the national and the local levels;
- limited publication and availability of nonsecret data as a result of poor reporting at the local level, no active policy towards publication and dissemination, and limited Internet use and access.

Often only general and aggregate official data are available, and then only for political decision makers, although specific local data is lacking or kept secret for those directly involved in and suffering from

---

[17] This list, consisting of six volumes and following the IUCN red list, was set up by the Biodiversity Working Group of the China Council for International Cooperation in Environment and Development, following three and a half years of work (2000–2003) carried out by more than one hundred scientists (see Wang Sung and Xie Yan, 2004; McBeath and Leng, 2006).

[18] In an analysis of the reliability of economic data and statistics, Holz (2003) found that especially at the higher, aggregate levels of the policy-making and bureaucratic systems, there is not much chance of deliberate falsification of statistical data. But the sheer variety of data that are collected and calculated by the lower echelons and sent to the central level give the National Bureau of Statistics a remarkable freedom in selecting which data best suit political purposes.

[19] Even the Chinese Academy of Sciences has to buy environmental data for research purposes (interview, division director CAS, November 2005).

environmental pollution. Consequently, local EPBs rely strongly on citizen complaints as monitoring data, and priorities for control and enforcement are more than incidentally set accordingly, instead of relying on their own monitoring and data collection programs (see later in this chapter).

In addition to, and partly as a consequence of, these problems, reliability and completeness of environmental data remain a major problem, in China as well as in Vietnam. The discussion on China's contributions to climate change via CO_2 emissions is telling in that respect.[20] But in China, more than in Vietnam, there is a clear tendency to further public disclosure and to improve and modernise the environmental monitoring system. For instance, the present eleventh five-year plan (2006–2010) has reserved RMB60 billion (around US$8 billion) for the entire environmental monitoring system. And, by 2008, China expects to have three additional satellites in the air for environmental protection and disaster control monitoring.[21]

E-government rather than e-governance

As noted earlier, China recently has become much more advanced than Vietnam in using the Internet, ICT and digital technologies[22] in

---

[20] At the turn of the millennium, a debate emerged on China's contribution to the greenhouse effect, against the background of the U.S. refusal to sign the Kyoto protocol. Sinton and Fridley (2001; 2003) and Chandler (2002) reported a decrease of 17 percent in China's greenhouse gas emissions between 1996 and 2000 (based on official Chinese energy statistics), the International Energy Agency estimated energy reduction to be 5 to 8 percent in that period, whereas the American Embassy in China claims a zero growth of energy use in China (http://www.usembassy-china.org.cn/sandt/energy_stats_web.htm, accessed November 2003). See, for recent data on CO_2 emissions, the World Resources Institute (http://earthtrends.wri.org).

[21] Originally, these were planned for 2006, but this seemed to be too optimistic. By the end of 2005, the complete proposal for environmental satellites was at the planning commission and fifteen staff members were continuously working on this plan at the EMC (interview deputy director EMC, November 2005).

[22] From 1993 to 2001, China and Vietnam had the highest average annual growth rates in the world on ICT spending, 27 percent and 26.5 percent, respectively. In 2004, China and Vietnam spent 4.4 percent and 2.4 percent of their GDP on ICT, respectively, and US$66 and US$20 per capita, respectively (the average for low-income countries is 4.2 percent of GDP; source: World Telecommunication Development Report 2006 database).

communicating and disclosing environmental information. Does that mean that we see e^2-governance emerging in China? And, if so, what are its similarities and differences with e-governance in OCED countries?

During the past one and a half decades, China has indeed invested significantly in e-government initiatives, and the annual budget for e-government is still increasing annually at a rate of around 40 percent (Yong, 2003: 83). Although perhaps in absolute figures the amount falls short of what is happening in OECD countries, when compared to other developing countries, the investments in e-government in China have been enormous.[23] According to Kluver (2005), the purpose of these massive investments in e-government in China is not so much related to empowerment of citizens or furthering democracy by increasing citizen participation, but can be related to three goals: (i) stabilisation and increased efficiency of bureaucratic procedures and – to a lesser extent – state-citizen interactions, also to relieve the rather obsolete bureaucratic structure; (ii) to reduce corruption, by providing means and techniques that enhance transparency, surveillance and control;[24] and (iii) combat the distortion of information, which seems to be inherent in hierarchical systems without transparency. This all comes together in the aim of strengthening the centre vis-à-vis lower level governmental agencies, enterprises and citizens. To a major extent, as also emphasised by former Premier Zhu Rongji, the various e-government projects (e.g., the so-called Golden projects, the Government Online project) focus rather on improving 'government' than on the 'e'.[25] Electronic government is seen as a project, through

[23] Quite surprisingly, the Global E-Government report of 2005 (West, 2005), ranks China fifth of all countries on e-government, using criteria of information availability, service delivery and public access. Vietnam, in contrast, ranked only seventy-seventh out of the more than 130 countries.

[24] This is done especially via the Government Online Project 2000 (available via http://www.gov.cn/govonlinereview). In September 2006, Chinese official news agencies reported that during the first eight months of 2006, US$15.6 billion of illegal state financial transactions was identified. This is all part of the offensive of Chinese leaders in 2006 to fight corruption, especially by enlarging transparency in decision making and by information disclosure.

[25] "We must emphasise the usage of the means of informationalisation, in order to strengthen the efficiency of government work. This will make the supervisory work of the government more serious, effective, increase the proximity to people, and raise the efficiency of service. It will make every level of government clean, industrious, pragmatic, and highly efficient" (speech of former Prime Minister Zhu Rongji in 2003, as quoted in Lagerkvist, 2005: 193).

which more efficient and modern bureaucratic government can be furthered (Lagerkvist, 2005).

Although most of the efforts of the Chinese state regarding the Internet have been on improving the bureaucracy, this is not really reflected in international studies on China and the Internet. The majority of the academic studies on the Internet in China focus on the government's attempt to control and restrict the Internet (see later in this chapter), and only few studies have concentrated on the large-scale investments, commitment and dedication of the Chinese state to develop the Internet and e-government: building digital infrastructure, mandating official use of the Internet and encouraging participation of citizens with digital technology (Kluver and Chen, 2003).[26]

Within the field of environmental governance, the use of digital technologies has greatly improved data collection and data availability. The Web site of SEPA, as well as many Web sites of provincial and local environmental protection bureaus, contains large amounts of environmental data on environmental investments, environmental quality, citizen complaints and sometimes even emissions. Also most of the relevant laws and regulations on the environment can be traced back on the Internet. Compared to the e-governance structures in OECD countries, however, there are hardly any possibilities for citizens to participate, to gain insight in the procedures of policy making and lawmaking and to forward with ideas and comments. The e-government initiatives on the environment are also one-way top-down initiatives, without any input from citizen discussion groups or NGOs. This means that these EPA Web sites can play only a minor role in issues of accountability, transparency, legitimacy and responsiveness. The National Informatization Steering Group (CCW Research, 2002) also noticed that the problem with Chinese government Web sites is that they are superb in providing texts or regulations and laws, but generally fall short of possibilities for citizens to interact with the government or provide a means for citizen consultation. It is here that significant differences with e-governance developments of OECD countries exist (Qui and Hachigian, 2004). Hence, with respect to China, e-government seems to blossom, whereas e-governance is largely absent.

[26] See Walton (2001), who explains the so-called Golden Shield Project in China. Usually Western multinationals (such as Silver Bullet, United Kingdom, and Verint Systems, United States) are active in selling equipment to carry out surveillance and control over Internet and mobile phoning in Vietnam and China.

3. Transitional state-market relations

With a strong record in command-and-control regulation and a poor history in public disclosure of environmental information, both transitional economies have relied mainly on state enforcement mechanisms to move industrial polluters into more sustainable development paths, with ambivalent successes at best. Although very general legal provisions for corporate environmental reporting and responsibility have been in place for some time, until recently these were hardly operationalised in concrete practices. However, recently, a number of initiatives can be noticed within China – and, to a lesser extent, Vietnam – to use informational dynamics in environmental governance. Some of these innovations are domestically organised, but most have a strong international dimension and are closely linked to the integration of these countries in the global economy and polity.

In China's State of the Environment Report 2000, public access to and disclosure of information was strongly pushed by SEPA. With assistance of the World Bank, and following successful experiments in other Asian developing countries (e.g., Indonesia, Philippines, South Korea[27]), at the turn of the millennium the Chinese SEPA developed its so-called Green-Watch program. The program aims to disclose the pollution record of industrial companies to the media and the public as well as pressure these companies into improving environmental performance. Two experiments in Zhenjiang (Jiangsu province) and Hohhot (Inner Mongolia) (cf. Wang et al., 2002) had different rates of successes. Following public ratings of the environmental performance of companies in these cities, a significant number of companies improved their performance before the next rating one year later. However, an evaluation in 2005 (Wanxin Li, 2006) illustrated continuing enthusiasm in applying disclosure programs in Zhenjiang but less so in Hohhot. The Green-Watch program has been further extended in Jiangsu province to an increasing number of municipalities and firms, although successes remain dependent on engaging governmental and industry participants, keeping levels of complexity low, full governmental support, availability of capacity at the local EPBs, and clear

---

[27] Indonesia's PROPER program and Philippines' disclosure program were developed in the mid-1990s (World Bank, 1999). South Korea has for a much longer time practiced the disclosure of polluting enterprises (Dasgupta et al., 2004).

communications to the media and public in terms of ratings.[28] Since the turn of the millennium, many environmental rating systems have been developed by local and provincial environmental authorities, for instance, in Shandong Province, Jiangsu Province, Hebei Province and Anhui Province. Similarly, the 2004 Cleaner Production Promotion Act has been instrumental in informational governance (cf. Guo Peiyuan, 2005; Mol and Liu, 2005). The law enables SEPA and local environmental protection bureaus to publicise environmental data of noncomplying companies in newspapers or on Web sites.[29] Early in 2007, SEPA published a 237-page list of more than six thousand industrial polluters on its Web site, requiring them to install automatic monitoring and control systems that are directly connected to local EPBs and plan for disclosure of this collected data. SEPA also issued regulations on environmental inspection and public disclosure of environmental performance for companies accessing – or refinancing on – the stock market.[30] And in May 2007 SEPA unveiled a new transparency rule requiring all government environmental agencies and polluting companies to

[28] To further stimulate public access to environmental information, the Professional Association for China's Environment had, for its 2005 annual conference, public access to information as the central theme. This conference was strongly pushed and subsidized by the World Bank and SEPA, the two main organizations behind the Green-Watch initiative. This program was later developed by SEPA into the Environmental Performance Information Disclosure programme.

[29] Since implementation of the Cleaner Production Promotion Law in 2004, many provinces used this law to force noncomplying companies to publish emission data (using article 31) and EPBs have publicly criticized poorly performing enterprises (based on article 17). For instance, the Guangdong Environmental Protection Bureau released a list of thirty-three companies that were required to accept cleaner production auditing and publish in the media details of their principle pollutants (Guo Peiyuan, 2005: 7–8). In 2004, Sichuan Province required 431 companies to disclose environmental information monthly. SEPA has detailed the relevant information disclosure articles of this law, including the kind of information that need to be disclosed and how it was to be disclosed, in the *Bulletin on Information Disclosure of Corporate Environmental Performance*, on 5 November 2003.

[30] *Regulations of Environmental Inspection on Companies Accessing or Refinancing on Stock Market*, on 16 June 2003. By this regulation, heavy polluting companies (e.g., metallurgy, chemistry, oil, coal, construction, paper and pulp, textile, leather, etc.) that want to access the stock market and listed companies that want to refinance on the stock market for investment should report emission levels to the EPBs. These EPBs should carry out site inspections to verify environmental compliance, disclose this verification report in the mass media, and report to SEPA and the China Securities Regulatory Commission.

disclose important environmental information to the public. The rule will take effect in May 2008 and is part of the wider *Regulations on Open Government Information* recently issued by the State Council to open up access to government information for ensuring greater official transparency nationwide.

Based on a similar idea of informational governance, HCMC environmental authorities in Vietnam have for several years used various blacklists and green lists, which mention the most polluting and the most environmentally sound companies, respectively.[31] Local authorities in HCMC and some other industrial centres try to use public disclosure of these blacklists to pressure the most severe industrial polluters, with very limited impact (Phung Thuy Phuong, 2002; Le Van Khoa, 2006). Green lists show equally limited incentives and pressure for environmental improvements. Authorities in the cities of HaNoi and DaNang have experimented with rating systems for polluting and green companies, be it with limited disclosure of poor performing company names to the public (Faucher, 2006). These list and rating systems are strongly pushed by bilateral and multilateral donor projects, and sometimes show major discontinuities. In addition, local environmental authorities are increasingly active to report in local and national newspapers on heavily polluting enterprises, building public and political pressure on these companies.[32] However, compared to the widespread, diverse and partly institutionalised Chinese initiatives, Vietnam's environmental disclosure programs and practices look rather marginal and bleak.

Both in China and in Vietnam, information disclosure by companies is strongly related to governmental pressure. Environmental authorities

---

[31] In 1994, the HCMC government produced its first blacklist of companies, in 1997 its second, together listing eighty-seven heavy polluting companies. A survey of these eighty-seven companies in 2001 (MPI, 2001) revealed that 90 percent of these companies did not see any impacts of these lists on their business (e.g., on sales, credit facilities), and 39 percent even believed that it was advantageous to be on the blacklist as it opened governmental and international donor support programs. The Green Book listed nineteen companies, which overall believed that the listing positively affected their profitability, reputation and future environmental performance (MPI, 2001).

[32] HEPA in HCMC has an agreement with one of the major local newspapers, *Saigon Giai Phong*, to fill one page a week, disclosing all kinds of information on major polluters, environmental problems around infrastructures and so on. Media reporting on poor performing companies in DaNang and HaNoi seems less frequent.

especially force companies to disclose environmental information on production processes and locations. And Chinese (and, to a much lesser extent, Vietnamese) environmental state agencies are starting to make such data available to the public via mass media and Web sites, instead of keeping information within the hands of SEPA or NEPA and their subsidiaries. In addition to authorities, other stakeholders (NGOs, investors, credit institutions, customers, citizens and consumers) hardly put pressure on domestic companies to release environmental information, nor do they use available information to move companies into greening. But this becomes different once international forces enter the stage, for instance, via joint ventures (cf. Guo Peiyuan, 2005), via international trade or via international standardisation, as will be illustrated in the next section on labelling and certification.

Labelling, certification and international markets

A clear field in which we can see information at work in greening production and consumption is in all kind of labelling and certification initiatives, both related to products and processes. Most labelling schemes do not involve any legal requirements but instead refer to voluntary standards that are communicated via symbols in order to address specific markets. Many of the product and process labelling and certification schemes that do make an impact on environmental performance in China and Vietnam are related to international markets, such as production process certificates of the International Standardization Organization (e.g., ISO 14001; Figure 10.2),[33] the international HACCP (Hazard Analysis Critical Control Point) program, Good Manufacturing Practices (GMP) schemes (also promoted domestically), and product-specific labels and information requirements by private international customers especially in OECD countries.[34] Information disclosure dynamics and informational governance is then part

---

[33] Cushing, McGray and Lu (2005) show that the regional spreading of these certifications parallel lines of economic development, where the major cities of Shanghai and Beijing and the rich provinces of Jiangsu and Guangdong house the majority of ISO 14001–certified firms.

[34] There are exceptions, such as the mandatory China energy efficiency label for products (in effect as of 1 March 2005 for fridges and air conditioners; see also McNeil and Hathaway, 2005), which is meant for domestic use. But the design and use of the label follows closely the European system of energy labelling of products (Council Directive 92/75/EEC), making it at the same time fit for exports of electric appliances.

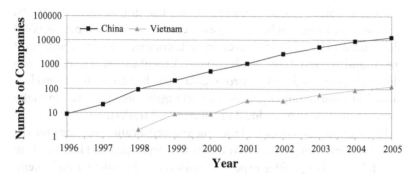

Figure 10.2. Companies with ISO 14001 certifications in China and Vietnam, 1996–2005 (*source:* http://www.iso.org/iso/).

of economic globalisation processes and/or an emerging global civil society. As an example, and in line with Chapter 7, we will elaborate on food labels and certifications.

In China, a significant number of domestic green, organic and healthy label programs have started in the field of food products (but also on energy conservation, water conservation and building materials).[35] With respect to the broad category of green food, the China Green Food Development Centre (established in 1992 under the Ministry of Agriculture) and the China Green Food Association (in 1996, also under MoA) have been active (with more than three thousand certified Green Food producers in 2002; Tracy, 2005), whereas for organic food the Organic Food Certification Centre (under MoA) and the Organic Food Development Centre of China (since 1994, under SEPA) carry out inspections and certifications of national standards and were present in over one-third of the provinces in 2005. Although a small domestic market does exist for especially green food products and to a lesser extent organic products (UNESCAP, 2002), most of the organic food production is for the international, Western markets, using well-recognized labelling and certification schemes.[36] The

[35] Cf. Guo, 2005; Qian Yi et al., 2001; UNESCAP, 2002; Tracy, 2005. See also the Web site of SEPA: http://www.sepa.gov.cn. Healthy food is food produced with a basic safety line to ensure health and food safety. Green food, introduced in 1990 by the Ministry of Agriculture, refers to 'safe' food produced according to strict standards of pesticides and fertilizers. Organic food production uses no pesticides and chemical fertilizers.

[36] The value of Chinese organic food exports in 2003 was about US$142 million, with more than one thousand farms and companies certified. The following

most important organic crops are coffee, tea, grains, nuts, dry fruits, rapeseed and sugar, whereas organic cotton is in development (Guo Peiyuan, 2005). China is increasingly becoming active in the field of green and organic food production and certification. Here, it is not domestic command-and-control regulations that green (food) production, but informational governance arrangements that link especially international markets to local producers and traders.

In Vietnam, such organic labelling and certification schemes for the domestic and international market are less developed. In HaNoi, 'clean vegetables'[37] (vegetables produced with limited pesticides and chemical fertilisers) are sold and some (tourist) restaurants advertise with organic products, but the domestic certification and verification procedures are not clearly developed. At the moment, the Department of Plant Protection (under the Ministry of Agriculture and Rural Development) is in charge of issuing certificates for 'safe vegetables'. However, what this organisation really does is give a (temporary) certificate for 'safe vegetable production cooperative' not for the product itself, because the department cannot be responsible for the quality of vegetables. A few years ago, the Department of Plant Protection did also start to patronise some villages in Donganh district (HaNoi) to produce safe vegetables with the brand name 'Bao Ha'. Some companies, such as Angiang Plant Project Service Company, launched their own vegetable brand names, in this case called 'Sao Viet'. However, these brands have hardly become known by domestic consumers in Vietnam. None of the vegetable producers/retailing chains have really gained the full trust of domestic consumers. In addition, no monitoring system exists in Vietnam that can help consumers to identify vegetable quality and/or trace back the vegetable to its origin. The major part of 'safe vegetables' produced by these producers are sold at open markets at normal prices. Another example of the underdeveloped environmental labelling is Dalat, often considered the most famous vegetable producing area in the South of Vietnam. Only recently, by the end of 2004, the Provincial

---

international certification organizations are active in China: BCS Öko-Garantie GmbH (Germany), Crop Improvement Association (OCIA) (USA), ECOCERT (France), Gesellschaft für Ressourcenschutz (GFRS) (Germany), IMO (Switzerland), Japan Organic and Natural Foods Association (JONA) (Japan), Quality Assurance International (QAI) (United States), the Soil Association (United Kingdom) and SKAL International (FAO 2005).

[37] As set by Decision 67-1998/QD-BNN-KHCN, issued by Ministry of Agriculture and Rural Development (MARD) on April 28, 1998.

Department of Tourism, Commerce, Science and Technology in cooperation with the Centre for Business and Commercial Enhancement
(based in HoChiMinh City) started thinking about branding vegetables
of Dalat.[38] Vietnam does produce organic food (such as organic cashew
nuts, tea, grapes and organic pineapple juice; it is aiming to enter major
crops such as organic coffee) for the international markets, especially
in the United States and Europe, using internationally recognised labels
and certification organisations (such as SKAL International). But the
institutional setting and capacity for a more coordinated take-off of
organic food production and export is poor. Labelling and certification
institutes, standards and technical support, knowledge, and specified
distribution and marketing channels are hardly developed compared
to China.

Other environmentally relevant information also can be found incidentally – for example, references to GMOs in China (Keeley, 2006) –
but in general the relevance of these kinds of information for domestic
consumers and producers is rather limited.

In all, the actual domestic consumption of and domestic pressure for
organic food is low in both countries, as a result of low incomes and
purchasing power, lack of knowledge and awareness of organic food
among broad segments of the population, lack of shops and retailers
willing to sell these products, confusion regarding the different labels,
standards and certification schemes (e.g., clean vegetables, green vegetables, organic vegetables, healthy food, as well as a number of private
labels), and a civil society that fails power and possibilities to press for
cleaner food production (UNESCAP, 2002; UNCTAD/WTO, 2007;
Thiers, 1999).

4. Transitional democracy? Civil society, public space and media control

China and Vietnam have been characterized for the past twenty years
by limited freedoms for the press and civil society in countervailing
the political elite and in criticising – among others – the disastrous
environmental consequences of a one-sided development path. We
have seen, however, that such public spaces are essential for effective

---

[38] See http://www.vneconomy.com.vn/vie/index.php?param=article&catid=
0909&id=041222143232, accessed September 2005.

informational governance. How far does the transitional process go in allowing a public space with an active civil society to access and use information in pushing for environmental reforms?

China has a very recent history of environmental NGOs and other social organisations that articulate and lobby for environmental interests and ideas of civil society amongst political and economic decision makers (see Qing and Vermeer, 1999; Ho, 2001; Yang, 2005: 51; Martens, 2006; Stalley and Yang, 2006). As the first national environmental NGO was only established in the mid-1990s, the history of this sector is rather short. For a long time, government-organised NGOs (GONGOs), such as the Beijing Environmental Protection Organisation and China Environment Fund, dominated the environmental 'civil society' sector.[39] They had, and still have, more freedom of registration and manoeuvre than independent NGOs, because of their close links with state agencies. Through closed networks with policy makers and their expert knowledge, these GONGOs articulate environmental interests and bring them into state and market institutions. In doing so, GONGOs play a role in bridging the gap between NGOs and civil society, on the one hand, and the state, on the other, thus "becoming an important, non-state arena for China's environmental politics" (Wu, 2002: 48). Recently, these GONGOs have gained more organisational, financial, and political independence and autonomy from the state and are (thus) evaluated more positively by Western scholars. At the same time, environmental NGOs are developing rapidly, although these NGOs remain embedded in the Chinese state. Figures on the number of NGOs are unreliable. Economy (2005) and Stalley and Yang (2006) estimate around two thousand registered NGOs and an equal number of unregistered ones, whereas Yang (2005: 51) provides more moderate (although rapidly increasing) numbers.[40] For the southern province of Guangdong, Ping Song (2005) counted seventy-three

[39] There are many kinds of GONGO, including foundations, education centres, research institutions and industry associations. They are able to play a major role in social and political developments because of their less restrictive institutional structure, their expertise and the personal connections with political elites.

[40] In China, NGOs need to be registered. Registration can be in different forms, but usually it means that these NGOs function under the umbrella of another, existing organisation (a governmental organisation, a university, a research institute or a private business). But industrial associations and professional societies are often also included as environmental NGOs, making numbers

formally registered provincial environmental NGOs in 2002, in addition to district-level, county-level and unregistered environmental NGOs. These NGOs, of which the majority are provincial or local ones, often are not very adversarial or confrontational, but, rather, are expert or awareness-raising organisations, such as Global Village. The 'political room' for a Western-style environmental movement still seems limited, but compared to a decade ago this room is expanding. In addition, Stalley and Yang (2006) show that there is also little interest from potential participants (in their case, university students) in joining and supporting environmental NGOs, resulting in small numbers of volunteers and supporters. In China, the contribution of environmental NGOs in pushing for environmental reform of the Chinese economy or polity has been marginal until now. International NGOs, such as Greenpeace and the WWF, have invested major efforts in further stimulating the environmental movement in China, with ambivalent successes.

In Vietnam, the situation of independent NGOs is worse. In addition to a few international environmental NGOs such as ENDA (which are only allowed if they cooperate with and not act against the government), Vietnam has only GONGOs and expert groups, the latter often related to universities. Registration of and governmental control over domestic NGOs or social organisations is more strict than in China. But in both countries, civil society is involved in various forms of informational governance, although often distinct from OECD countries. We will first look at complaint systems and subsequently focus on the conventional and the new media.

Civil society and informational pressures

Together with economic liberalisation, decentralisation of decision making, and experiments with local democratisation, one can witness a growing pressure of – often unorganised – citizens on local (environmental) authorities to reduce environmental pollution. Individual citizens, communities, local organisations and the media have been encouraged to come with independent information on pollution, among others, via various complaints systems. Dasgupta and Wheeler

difficult to compare with OECD country statistics. Registrations in Vietnam differ from this.

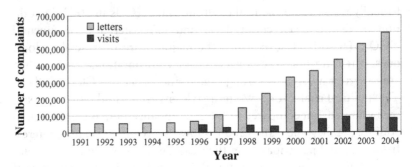

Figure 10.3. Environmental complaints by letters and visits to Chinese EPBs, 1991–2004 (China Environment Statistical Reports).

(1996) estimated that in China local and provincial authorities responded to more than 130,000 complaints annually in the period 1991–
1993, whereas Chinese data show lower figures for these years, but a
sharp increase from the mid-1990s onwards to six hundred thousand
in 2004 (Figure 10.3). In most Chinese cities and towns, systems of
complaints and telephone hotlines have been installed, albeit with vary-
ing levels of use and effect.[41] In China, this system of complaints and
the growing attention the (state-owned and -controlled) media pay to
environmental pollution and environmental mismanagement are more
important than NGOs in influencing economic and political decision
makers. This emerging environmental consciousness is, to some extent,
also reflected in the various polls that have been held (Stockholm Environment Institute, 2002; see Lee, 2005 for an overview of these polls).

   In Vietnam, similar dynamics exist but – in line with the country's hesitations to give civil society more space – less frequent,
less facilitated via, for instance, hotlines, and not systematically

---

[41] Dasgupta and Wheeler (1996) show that the average number of environmental
complaints of major cities and provinces in one year ranges from 55.0 per
100,000 inhabitants in Shanghai to 1.7 per 100,000 inhabitants for Gansu
province. In most provinces, EPBs responded to more than 80 percent of these
(telephone calls, letters and face-to-face visits) complaints, as also follows from
the Chinese statistics (China Environment Statistical reports, 1991–2004). See
also Brettell (2004) for a full and detailed overview of complaints in all
Chinese provinces and major cities.

reported.[42] Studies of O'Rourke (2004), DiGregorio and colleagues (2003) and Phung Thuy Phuong and Mol (2004) register the importance of complaint systems in Vietnam, evolving into community-driven regulation and governance on the environment. In such forms of community-driven governance, informational resources do play a significant role. Sometimes, information is spread widely through the various media to build up pressure on the political powers to act; sometimes information is channelled via informal and formal networks to higher party and state echelon, who act than in conventional – command-and-control – ways.

Media and public space

The 'old' media (television, radio, newspapers) are still largely owned, monopolised and controlled by the Chinese and Vietnamese state authorities, be it not to the same extent in the two countries, leading to somewhat distinct roles of these media in informational governance of the environment in the two countries.[43] In both countries, the conventional media have long been used primarily for propaganda and government-controlled information dissemination. In an interesting analysis, De Burgh (2003) explains the major changes that took place in the 'old' media and journalism practices in China. Part of the old media has been given economic independence and competition emerges between newspapers (but not yet on television), resulting in the need to get a major share of their funding from advertisements (up to 60 percent in newspapers) and more attention for consumer preferences.[44] Media staff is increasingly recruited outside party control and financial incentives are used to attract good professionals, including

[42] The 1992 constitution articulates the right of citizens to lodge complaints, and the 1998 Law on Complaints and Denunciations elaborates this constitutional right. Decree 67/1999/ND-CP further details the implementation of this law.

[43] In the World Resources Institute indices on environmental institutions and governance, the two countries score equally on most of the governance indices, such as Transparency, Level of Freedom, Civil Liberties and Press Freedom (http://earthtrends.wri.org).

[44] In 2004, there were some twenty-one hundred newspapers in China, some of them organised in groups (data from the World Association of Newspapers: http://www.wan-press.org/). Few are national newspapers. Most are either regionally focussed (provincial, major cities), for special target groups

journalists (Fang, 2002). In addition, state control has been relaxed somewhat and freedom on reporting has increased, although control remains especially over the more sensitive issues.[45] In entering these nation-states, the global media, such as satellite TV, are still strongly dependent on the Chinese (and Vietnamese) authorities, and can only broadcast and sell under specified conditions for certain markets (Sparks, 2005: 39–40).[46]

Environmental issues increasingly are seen as nonsensitive issues, which facilitates the revealing of environmental accidents, disasters and routine pollutions in China.[47] Research by Li Junhui (2005) and Hu Kanping and Yu Xiaogang (2005) on newspaper reporting on dams, arguably one of the more sensitive environmental issues in China, and of Wang (2005) on industrial parks illustrate that especially local newspapers feel the pressure of local state authorities to refrain from reporting critically, whereas critical environmental reporting is more allowed at a national level. Nationally, bans on reporting emerge when minority issues and national (security and economic) interests are involved.[48] Based on several sources, Yang (2005) concludes that environmental

[e.g., Youth League) or published by special organisations and agencies (such as, for instance, SEPA).

[45] As Fang (2002) and Li Junhui (2005) explain, Chinese news media are regulated and controlled via five mechanisms, of which the first two are the most important: government administrative system; Party committees; the legal system; social surveillance of other parties and social groups; (self-)regulations from associations in the news industry. Xinhua is China's leading integrated news company, with more than eighty-four hundred employees and a president that has the rank of minister (Battistella, 2005). It is state-owned and tightly controlled, especially since the 1989 Tiananmen Square event. Its journalists are tightly controlled and will only write critically when allowed. In September 2006, Xinhua announced further restrictions for the operations of foreign news agencies in China, as well as for the spreading of news in China (http://www.rsf.org/).

[46] See Whittemore (1990) on how CNN managed to report globally on the 1989 Tiananmen Square event, before the Chinese authorities eventually gained news control.

[47] A content analysis by the Chinese NGO Friends of Nature among tens of national, provincial and local newspapers showed that coverage of environmental items in these newspapers especially increased in the second part of the 1990s (Friends of Nature, 2000). This does not seem different in the other traditional media.

[48] Several Chinese journalists have been jailed following too critical reporting, either within their country or in foreign media (e.g., Zhao Changqing in 2003; Ren Ziyuan and Li Yuanlong in 2006).

NGOs and their campaigns have been treated favourably in the Chinese newspapers from their emergence in the mid-1990s onwards. There are close ties between the Chinese conventional media (newspapers, radio, television) and environmental NGOs (Hu Kanping, 2001; Hu Kanping and Yu Xiaogang, 2005; Xie and Mol, 2006; Xie, 2007), and more than incidentally green NGOs are led by (former) professional journalists. These close ties also come from the fact that environmental NGOs are a source of news, and pollution victims and environmental NGOs need the media to build up pressure.[49] Although this freedom has caused larger uncertainty among journalists and media decision makers of what is and what is not allowed, by the same token most journalists and media are less and less willing to accept simple top-down Party directions. Today, Chinese media serves two masters: the Party and the market (Hong Lui, 1998; Latham, 2000). The media seem to constantly experiment with the limits of what is allowed by the party (Li Junhui, 2005), whereas the limits prove to be a moving target. Especially since the SARS (Severe Acute Respiratory Syndrome) epidemic in 2003, the room for revealing environmental information has expanded.[50] The emerging 'investigative journalism' also focuses on scrutinising authority, although journalists and media seldom touch on 'Chinese leaders in action' or challenge the (local) state legitimacy, unless it is allowed from above. In that sense, transparency has increased in environmental governance, but governmental control is still felt and transparency in environmental governance through these old media is far from routine matter.

In Vietnam, the 'old' media also have been removed from their subsidies but are still owned by the government (including TV, radio and the five hundred to six hundred dailies and periodicals). Compared to China, there is a stronger control by and work in service of the Party and government. Aggressive editors and journalists who push too hard at the boundaries of censorship have been removed and

---

[49] Shang Hongbo (2004) studied how pollution victims in four cases of industrial pollution in different parts of China used the media to build up pressure once the (local and national) authorities did not prove to be receptive to their complaints. Next to these strategies of informational governance, litigation, protesting and mediation with the polluters are other strategies used by these victims.

[50] Interview, Tsinghua University professor, October 2005; interview, deputy director EMC, November 2005; interview, Chinese Academy of Sciences division head, November 2005.

replaced by more 'reliable' alternatives. Newspapers and periodicals are closed when writing too critically on the state or the Party. Market competition is still limited in Vietnam's newspaper branch and television is strongly in hands of the state. Investigative journalism is hardly known. Although critical reporting on issues such as corruption, pollution and crime happens, they remain in the official party line and never touch on the leaders and party in charge. In that sense, compared to China, the conditions for informational governance through these conventional media are less favourable in contemporary Vietnam. Independent NGOs seem to have also a more difficult position in Vietnam than in contemporary China, with few international but hardly any domestic environmental NGOs active in the mediascape.[51]

Compared to Vietnam, the 'old' media in China are much more active in collecting, publishing and broadcasting environmental information, misbehaviour and protests. The national governmental authorities especially see information disclosure, media coverage and (un)organised civil society complaints as an important correction mechanism of the local priorities towards a one-dimensional development path. Openness and nonstate information disclosure on local environmental practices are important channels for the central state to supplement vertical information flows through state structures. Although information is still strongly controlled by the government and company specific information about emissions is still either nonexistent or not publicly available, the Chinese authorities allow more openness and criticism to correct environmental misbehaviour and poor functioning of environmental authorities. With this, in China – more than in Vietnam – environmental information starts to become more than just an instrument in the hands of the government and party. Environmental NGOs, industrialists, different state fractions and individuals use – or try to use – environmental information in their struggles. But, compared to both the dominancy of conventional modes of governance

---

[51] It is surprising to see the high ranking of Vietnam in several lists of numbers of (environmental) NGOs, such as the one from the World Values Survey 1999–2002 (indicating that 7.6 percent of the Vietnamese population is a member of an environmental NGO; see, for instance, Dalton, 2005) and the World Resources Institute database (http://earthtrends.wri.org), indicating ten international NGOs per million inhabitants. One wonders what the definition of NGO is in these databases.

in China and Vietnam (where the state uses its authoritative resources to command), and the informational governance practices in OECD countries, these first sprouts of Chinese and Vietnamese informational governance are rather bleak. In an interesting analysis of environmental governance towards GMOs, Keeley (2006) identifies what he calls a proto-public sphere, in which there are increasing possibilities and civil society interests to discuss and debate on such controversial and economically important developments as GMOs, but in which, at the same time, most relevant information with respect to tests, risk assessments and environmental consequences remains secret.

The new media

The new media – in contrast – proves more difficult to be controlled by the transitional state, let alone owned and monopolised. Wilson (2004: 223–97) extensively analyses the efforts in China to both stimulate ICT as a key motor for technological development (with annual sectoral growth figures of 25 percent) and control the Internet in ways not too dissimilar from the old media.[52] In China, the temporary Internet restriction issued in 2000 has been further tightened with a new set of policies and measures in 2005. This is not different in Vietnam, although in that country state regulatory control seems tighter at the time of writing.[53] Both states have their policies and efforts to try to remain in control of Internet, for instance, via monitoring Internet use by consumers, requiring registration at local security agencies, limiting links or gateways between national and international networks,

---

[52] As most developing and industrialising countries, both China and Vietnam are eager not to lag behind – or even to take a competitive advantage over – other industrialised nations. Regions and countries at the margins of the global flow of financial and economic capital know what it is like to be delinked from such global developments, networks and flows. Hence, we see in most developing countries active plans, policies and investments on information technologies, contributing to a digital race in cyberspace (cf. Wilhelm, 2004).

[53] Decree No. 55/2001/ND-CP is the main regulatory framework for the management, provision and use of Internet services in Vietnam. Although this decree has quite strict regulations on Internet service provision and Internet access, implementation seems not that strict. But, compared to China, ICT is still largely in hands of state agencies (see also GIPI Vietnam [2004], *Promoting Internet Policy and Regulatory Reform in Vietnam*, The European Union's Asia IT&C program, Global Internet Policy Initiative Vietnam: HaNoi [http://www.internetpolicy.net/about/20040300vietnam.pdf]).

temporary bans on Internet cafés, placing of cameras and requesting identification in Internet cafés (by more than five hundred Internet caféwatchers in Beijing in 2006), closing of Web sites, limiting access to and production of news sites and weblogs, blocking access to 'undesirable' Web sites, intimidating actual and potential users, restrictive policies towards Internet service providers and jailing Internet activists.[54] The Chinese government is even able to condition major international Internet companies such as Yahoo, Microsoft, eBay, Skype and Google, which all seem willing to accept all kind of restrictions in entering this giant market.[55] In China, the combination of all these restrictions is often labelled 'the Great Fire Wall'. At the same time, in both states, these restrictions are to a significant extent subverted and are not always that effective. It seems that in China and Vietnam, the controlled liberalisation of the Internet in the 1990s made place for a more free development of the sector at the turn of the millennium, but is anno 2006 again the battle ground between liberalisation and state control. Chinese and Vietnamese citizens and organisations employ various tactics to gain and sustain freer access to the Internet (cf. Wilson, 2004: 272), as they do with other restrictive policies. But major parts of the Internet remain in state hands and state control continues to be in place. China and Vietnam were also among the fifteen countries who plead – in vain – for more state control on the Internet during the November 2005 World Summit on the Information Society in Tunis,

[54] See Sinclair, 2002; Wilson, 2004; Downing, 2005; hrw.org (Human Rights Watch). The Chinese government was especially effective in hacking the Internet activities and Web sites of Falun Gong in 1999. With that, the government has been quite successful in controling Falun Gong Internet communications within its frontiers, but less so those coming from outside mainland China. In 2006, several Internet activists were still in prison in both countries (see the annual reports on both countries by Reporters sans Frontières at http://www.rsf.org/).

[55] Yahoo cooperated with the Chinese authorities in 2005 to track down and prosecute journalist Shi Tao (who was sentenced to ten years), and Google accepted limiting the results from its search machine in China, for instance, in order not to let Falun Gong, Taiwan, and Dalai Lama emerge in search results (see the Web sites of Reporters sans Frontières and Human Right Watch, for more cases like these; see also Goldsmith and Wu, 2006). Early 2007, Microsoft, Google, Yahoo, and Vodafone were discussing with several NGOs (including Human Right Watch and Reporters sans Frontières) an Internet code of conduct to protect free speech and privacy of Web users, following these events and allegations.

by strongly pushing for a new UN organization to manage the Internet instead of the U.S.-based nonprofit organisation Internet Corporation for Assigned Names and Numbers (ICANN).[56]

Especially in the urban centres, and more in (eastern) China than in Vietnam (Katsuno, 2005), Internet access has become widespread and relatively easy (compared to five years ago; cf. Figure 10.1). In 2003, China ranked first in the world in number of mobile phones and second in the number of Internet users (Harwitt, 2004). Digital environmental information, both domestic and international, has hardly been restricted by the state, especially not when controversies are at stake. During the famous 2005 Harbin disaster, when an accident at a chemical plant polluted the Songhua River in northeast China and stopped the drinking water supply in Harbin city for several days, it was people on the Internet who mobilised public opinion and said that there was something more at stake than a routine control of the drinking water supply as authorities tried to let the public believe (cf. Box 10.1). More generally, Chinese environmental NGOs have been quick to use the Internet, also because of the political restrictions in the other media in China. More than half of the environmental NGOs in China have set up Web sites with environmental information, bulletin boards and Internet campaigns. Some NGOs, such as the Green-web and Greener Beijing, operate only through the Web and are unregistered.[57] They publicise environmental information, set up discussion groups, mobilise volunteers, organise activities and campaigns[58] and catalyse offline campaigns. From a survey among urban grassroots organisations, Yang (2007) concluded that their Internet capacity is still at a low level and that especially the young organisations make active use of the Internet for publicity work, information dissemination and networking with fellow organisations, resulting in "a web of civic associations in China". In Vietnam, such national environmental Internet activism is still lacking.

[56] At this summit, Reporters sans Frontières listed these two countries among the fifteen "black holes of the Web" and "enemies of the Internet".

[57] Web-based groups can escape the formal regulations on registration. In a useful overview, Yang (2005: 50) distinguished seven different types of environmental NGOs in China, of which Web-based groups is one category.

[58] For example, Yang (2005: 63–64) reports on an online campaign in 2002 organized by Green-web, which successfully stopped the building of an entertainment complex that threatened a wetland.

Box 10.1 Harbin, China (Carter and Mol, 2006)

The environmental disaster in the Songhua River, centred on the city of Harbin in the northeastern province of Heilongjiang in November 2005, provides an illustration of the current trends that characterise China's system of environmental governance. At first sight, the Harbin disaster looked like a classic example of an industrial environmental accident that could happen anywhere, when old facilities, poor risk management and the limited environmental capacities of state and private agencies come together. An explosion at a large PetroChina chemical factory at Jilin on 13 November 2005 released a huge spill (more than one hundred tonnes) of highly toxic benzene into a major river, threatening the water supplies and river-based economic activities not only of various Chinese cities (among which the metropolis of Harbin 200 km downstream) and villages along the river, but also towns downstream on the Russian border. Only a decade ago, such accidents might have received little publicity in China or have been reported only as a successful example of disaster management by the local authorities weeks after the accident. Now, however, the old and new media coverage in China focussed not only in great depth on the disaster itself, but even more on the attempt by PetroChina, local officials and the national SEPA to conceal details of the pollution threat and then to release misleading information about it. Moreover, citizens held the company as well as their local government directly responsible and accountable for both its inadequate response to the incident (particularly its failure to safeguard water supplies) and for the attempt to cover up the disaster. And, at the national level, politicians held SEPA responsible and accountable for misinformation and cover-up, leading to the resignation of the SEPA minister, Xie Zhenhua. The rapid spread of the news to the wider world, especially via the Internet, put additional pressure on Chinese leaders to act directly and forcefully, and to provide full information (the order to go public with the information is said to have come directly from the State Council in Beijing). This example has all the ingredients of a mix of emerging informational governance with conventional Chinese state policy-making routines. Although the knee-jerk official instinct in China is still to suppress and conceal bad news and to follow state and party logics in solving the environmental crisis, the informational dynamics, China's integration in the global networks (e.g., the wide coverage of this disaster in all the major global news agencies and media) and the relative autonomy of the various Chinese media, all forced another form of governance with public accountability, disclosure of information and independent scientific scrutiny (by the leading Tsinhgua professor Jining Chen). Part of China's media has been particularly open and tough on the government and party reaction to the disaster, among which are the *China Youth Daily* and *China Newsweek*, although official Party instructions aimed at preventing too much news coverage and in-depth investigation.

Although this Harbin case was soon followed by reports in China and abroad of similar disasters of industrial water pollution and accidents and environmental governance innovations to deal with that, at the same time numerous other environmental irregularities and disasters continued, outside the domestic and global informational flows and pressures, and unable to activate the networks of power inside and outside China. Although the Harbin disaster gives us – also under conditions of poor information and information control – hope for the powers of informational governance, at the same time it shows that informational governance has its limitations.

The new media are also a vital source of environmental information, especially in these countries where information has been hard to get for such a long time. Environmental groups, students, scientists, concerned citizens and others have now access to a variety of information: on local and national facts and figures of environmental pollution, on environmental regulations and policies in other countries, on environmental movements in democracies and so on. In addition, they are able to get into easy and frequent contact with colleagues worldwide. In that sense, the new media have done much more in creating a knowledgeable and active civil society than the efforts of a limited number of international NGOs that have landed the past ten years in China and Vietnam. But one still has to be cautious not to overestimate the contributions of Internet to environmental democracy. It has been especially the urban intelligentsia and the organizations they work with, who have seen their political possibilities and empowerment increasing through the Internet, and not so much the larger share of the population. The distinction between those with access to the new modes of communication and information and those who have no access is large. The digital divide in China and Vietnam runs along economic and geographical lines. The eastern and southeastern areas in China (Katsuno, 2005) and the key economic zones around HoChiMinh City and HaNoi in Vietnam have the largest share of Internet connections, whereas – notwithstanding governmental policies and international development funding and projects – rural areas have much less access.

5. Conclusion: informational governance in *status nascendi*

It becomes clear that informational governance in China and Vietnam is at best in development, and far from matured. Although in China several sprouts of informational governance could be identified, Vietnam largely fails the institutional conditions for such a development at the moment: limitations on the development of an independent civil society, media and public space; an economic sector whose interests are still strongly intertwined with the state sector; a small part of society connected to the World Wide Web; and a poor state capacity in information collection, handling and dissemination make the conditions for informational governance in Vietnam meagre. Although the institutional conditions for informational governance in China look much more favourable, compared to the OECD countries in the European and North American continents China's information and media-scape is underdeveloped for informational governance. A significant part of informational pressure in the field of environment originates then from outside, from the global economy and international assistance programs that play an increasing role in Vietnam (and less from global civil society representatives that find a hard time invading the country). But then, of course, most informational governance practices that can be witnessed are not equally spread over the country. We find them in the economic hubs or the sites that receive strong international support and interventions, and less in the rural periphery and small towns.

Again, as stated in the introduction to this chapter, China and Vietnam cannot be taken as representative for the developing world. They are much too specific for that. But this chapter has helped us to understand the relevance of informational governance outside the core nodes that make up the global network society. Many of the dynamics that speed up new modes of environmental governance in OECD countries are only marginally relevant for countries without a vibrant public space and civil society, without a widespread dissemination of and access to free information and the Internet, with large groups of companies that are not too afraid of losing their reputational capital as they are not very 'visual' in the global economy, and with state authorities that lack capacity, means, fear for legitimation crises, and the desire to maximise environmental information disclosure on all pollution frontiers.

PART III

Conclusion

11 | *Balancing informational perspectives*

1. Introduction

In its 2005 report *Towards Knowledge Societies,* the United Nations Education, Scientific and Cultural Organisation (UNESCO) discusses to what extent the emerging knowledge societies are risk societies. UNESCO concludes that knowledge societies are rather the answer to – than the cause of – risk societies, as they constitute one of the most effective means of dealing with new complexities and risk. But, by the same token, UNESCO accurately describes – in their own words – the challenging issues at stake in where knowledge societies are confronted with risks and unsustainable development (UNESCO, 2005: 142):

The concept of knowledge societies and the central role played by networks correspond very precisely to these new requirements of collective action, capable of mobilizing all the resources of governance and science in real time as well as in the long term, with a pluralistic international approach in mind. [. . .] In the fields of sustainable development, environment protection and global health, the complexity of the data and the stakes involved exclude any possibility of a single response or a unique viewpoint, particularly where experts are uncertain when confronted by a new issue. The need here is to institutionalize, as it were, the fact that any question on a global scale is, initially in any case, too complex to command unanimity, even in the scientific world.

This diverts not too much from the central subject and analysis in this volume. Against the background of intensifying globalisation processes and an information revolution, both of which we have most likely only witnessed the early phases, this book has explored the changes, challenges and innovations in environmental governance. The core assumption behind this exploration was that core transformation processes of the modern order, such as globalisation and the information

275

revolution, cannot but affect the way modern societies deal with the environmental challenges they are confronted with. Or, to formulate it in one question: what does environmental governance look like in the Information Age? Environmentalists, environmental authorities and environmental scholars have not been very reflexive until now on what the core social transformations of the Information Age mean for their approach and understanding of environmental governance, although their actions, activities and investigations do already – marginally or more fully – echo the changing conditions under which environmental challenges have to be addressed tomorrow. In that sense, the UNESCO (2005) report has been exceptional in analysing how the advancement of sustainable development will be affected and changed as a result of the emergence of knowledge societies.

I have used the concept *informational governance* as a common denominator for the new environmental governance practices and arrangements, which emerge under conditions of the Information Age. This final chapter aims to make up the balance and formulate conclusions on informational governance, along several lines. First, I will summarise the core idea of informational governance and indicate the main advantage of understanding current developments in environmental governance as informational. Second, I aim to balance informational governance against conventional forms of environmental governance. Do we witness something new, how present are these new forms of governance, and how do they relate both to the conventional forms of environmental governance that are so strongly based on nation-states with their legal resources and sound science and to the new forms of (multilevel, multiactor) governance that have recently dominated the political sciences and public administration literature? Third, the variations in sprouts of informational governance around the globe will be emphasised, balancing the idea of a development towards a universal new mode of environmental governance worldwide. Fourth, an assessment is made of informational governance in terms of its environmental relevance (or effectiveness, conventional scholars would say) and its consequence for issues of democracy, new power balances, and the inclusion and exclusion of various actors and interests related to environmental reforms. Who are the winners and losers? Finally, a new agenda is formulated, based on these assessments: an agenda both for environmental governance under conditions of a global network society and for further research.

2. Informational governance: what is it?

In exploring to what extent and in which way the joint forces of the information revolution and globalisation will and do affect environmental governance I have introduced the label *informational governance*. Informational governance refers not so much to the fact that information is important for addressing and dealing with environmental challenges, as that has always been the case ever since modern states started to develop and implement their environmental activities and programs. Rather, the concept implies that for understanding the current innovations and changes in environmental governance we have to concentrate on the centripetal movement of informational processes, informational resources and informational politics. It is the production, the processing, the use and the flow of, as well as the access to and the control over, information that is increasingly becoming vital in environmental governance practices and institutions. Thus, the strategies, actions and coalitions of actors in environmental politics and governance, as well as the formation, design and functioning of institutions for environmental governance, can no longer be understood without focusing on information and knowledge. Information and knowledge are becoming key resources in environmental politics, the sites and spaces of environmental controversy relocate to information, and the motivations and sources for changing unsustainable behaviour are increasingly informational. The notion of informational governance on the environment enables us to understand and assess the significance of reputational capital of companies, of legitimacy capital of environmental NGOs, of transparency and disclosure policies, of new environmental monitoring arrangements, and the digital turn in the media – to name but a few – for environmental governance. Moreover, rather than noticing these new trends as relatively unrelated developments in environmental politics, informational governance brings these coherently together under one common denominator.

This centripetal movement of informational processes in today's environmental governance is not an autonomous or endogenous process that unfolds within the field of environmental governance. It should not be understood as just the logical answer to the state and governance shortcomings and failures in environmental policies. In the foregoing chapters, I have shown that it is closely connected with, and cannot be understood without paying attention to, a number of other key social dynamics and processes: globalisation processes, the changing sovereignty of nation-states (both internally and externally), the growing uncertainties connected to the disenchantment with science, and various technological developments. This close connection to these wider social developments makes that informational governance more than just a voluntary choice of policy makers or governance actors. It is structurally embedded and increasingly institutionalised within wider developments of global modernity and, as such, has some permanency. At the same time, informational governance is also not just a product of a technological revolution that has been so central for some of the Information Society scholars. Informational governance is as much logically linked to time-space compression, decreasing state authority and capabilities in environmental regulation, and growing complexities and uncertainties, as it is to technological transformation and change.

Although being structurally embedded in wider social developments, informational governance is not determined by these structures and developments. As has become evident from this book, informational governance is far from a well-crystallised uniform model, which unfolds in a similar way at different locations around the globe and on which governance actors can have no formative influence. Informational governance is still very much in the making; it takes different forms and shapes in different sectors and societies; it is part of and shaped by conflicts and struggles; and its relevance for environmental reform is certainly not equal in every corner of the global network society.

The focus on informational processes and informational resources in (environmental) governance contributes to a better understanding of the widely noticed shift in governance with respect to the diversification in steering modes, actors, and level-interdependencies. Conventional (regulatory and economic) modes of governance, dominated by state authorities, are opening up and diversifying through informational processes and resources. There is no one-to-one and/or causal relation between the emergence of new modes of governance and the growing centrality of informational processes and resources; but by including these latter developments into the governance literature, we gain insights in how and why state authorities and institutions see their power diffusing to other actors and arrangements, and what the (power) basis is for these new governance arrangements and actors.

3. Continuities and discontinuities

In evaluating the various contributions on the politics and culture in the Information Age, and in debate with Castells's ideas on informational politics, Webster (2001a) makes a strong argument against exaggerating the newness of politics in the Information Age. The empirical evidence of contemporary politics, not in the least those engaging environmental movements, shows only incidentally major transformations that can be interpreted in line with ideas of the Information Age. In elaborating on that, Webster warns us of the dangers of *presentism* and makes a strong case for noticing the continuities of the past: capitalism is still there today and much of the developments we see today can be interpreted as (a reaction or protest to) capitalism. Irrespective of the significant changes we are undergoing in politics, in movements, and in economics, these features persist until today.

Although stressing continuities and downplaying any discontinuities are common reactions toward all historical claims that a new era dawns, they should be taken seriously. Informational governance of the environment does not suddenly take over conventional forms of policy making and governance, strongly based on regulatory resources and state authority. Networks and informational flows have not made state and societies (completely) irrelevant categories for understanding current forms, modes and dynamics of environmental governance. Laws and regulations, largely sovereign states in an international order, and authorities that still have a monopoly of using state power remain crucial elements in contemporary attempts to govern sustainability, whether successful or not. It should be acknowledged that (sub)national environmental authorities continue to play a significant role in national and international environmental agendas, be it to a different extent and in different ways depending on the issues at stake and the locations in the global network society. Thus, nation-states, national societies, regulatory resources and sovereignty are still relevant concepts, as 'continuity scholars' claim. And it is too early to delegate these as 'zombie' concepts (Beck, 2004) to the past.

Although relevant for understanding continuities, these conventional concepts and interpretative traditions are no longer sufficient to understand current dynamics, innovations, and transitions in environmental governance. When relying solely on these conventional concepts and interpretations, we are unable to understand the dynamics and transformations in how hybrid arrangements blur conventional classifications of states, markets and civil society in governing environmental challenges. Equally, we lack the tools to grasp the changing position of conventional regulatory resources and actors, and the emergence and strength of new resources and arrangements that are so crucially tight up with information. Why are state actors on an increasing number of issues unwilling, unable or hesitant to use their regulatory power to push the sustainability agenda in a globalising world? Why are all kind of actors so preoccupied and worried with information disclosure, labelling and certifications systems, reputation and legitimacy, transparency and accountability, the old and new media, tracking and tracing, and monitoring among others? We cannot understand this by focusing on state command-and-control politics, relying on economic cost arguments for polluters and/or highlighting the necessary information preconditions for regulatory actions by state authorities. It is exactly the interpretative shortcomings of the conventional paradigm in understanding these contemporary changes and innovations in environmental governance that motivated me – and others – to search for new governance frames.

If indeed it becomes essential to include informational processes into the centre of our analysis of environmental governance, this does not necessarily means a change for the better. This can be seen as a second claim of continuity scholars: although perhaps the modes of environmental governance change, the success and impact of this new mode or form of governance is not any better than conventional systems of environmental regulation that were so strongly related to notions of state failure. Some scholars even claim that these new fashions and tendencies only weaken environmental governance, as the strong position that the environmental state has built up over the past three decades, becomes now seriously undermined. If formulated in this way, the conclusion of continuity scholars might be largely correct; but at the same time their analysis misses depth and understanding. Webster (2001a), Schnaiberg and colleagues (2002), and, to some extent, also Paehlke (2003), among others, are all correct in claiming that we are still dealing with a (global) capitalist economic order, which forms the setting for environmental governance.[1] And in such a setting environmental

---

[1] Through their study of car advertising, Paterson and Dalby (2006) provide a recent example on how information and symbolism works towards what they

governance faces structural problems in dealing with, 'regulating' and reforming powerful profit-driven private agents, organised in a market economy. But, by stressing in such a general way continuities, one easily misses the discontinuities and transformations, for instance, in how capitalism is organised and functioning today. Under conditions of globalisation and informationalisation, with a growing centrality of old and new media dominating the public sphere, the importance of information control (e.g., the American NSA with forty thousand staff members; the increasing amount of satellites; the growing struggle over Internet traffic control), decreasing nation-state power, growing system complexities and the high mobilities, contemporary 'capitalism' is fundamentally different from three decades ago. These transformations – that is, within a largely capitalist order – call for, trigger, force and determine new modes of – in our case – environmental governance. And although one could argue that contemporary modes of environmental governance do not fundamentally score better than the state failures of the 1970s and 1980s, such a conclusion misses any understanding of what is happening today.[2] Positions of actors, power balances, coalitions, resource dependencies, the rules of the 'environmental game', effective strategies and so on are changing rapidly around environmental controversies. If we want to understand current successes and failures, current strategies and discourses, current power balances and inequalities, current inclusions and exclusions around defining and handling environmental flows, we have to move beyond general continuities and constantly look for and interpret specific dynamics, changes, innovations and – also – continuities. The new modes and forms of environmental governance might not necessarily be better, nor necessarily be worse; neither more successful nor entailing larger failures; they should not be celebrated nor be condemned. But these new dynamics are 'fitting' the new conditions of our modern order and have not emerged accidentally. Before any celebration

call imperial politics of environmental degradation, contrasting and challenging our informational focus on environmental reform.

[2] An interesting question would be whether the conventional system of nation-state based regulatory environmental governance from the 1970s and 1980s would be more successful in dealing with today's environmental challenges in a global network society than the emerging informational mode of governance. There are good arguments to doubt how well the conventional regulatory environmental governance model would fit today's complexities.

or condemnation, one should first of all understand the logics and backgrounds of (the emergence and functioning of) these innovations and changes in environmental governance.

The scholars working on (new) modes of governance, as were introduced in Chapter 1 and at various other places throughout the book, would to a significant degree agree with this discontinuity analysis. This literature on shifts in (environmental) governance has pointed at the discontinuities in modes of governance following wider developments in society, paying special attention to the inclusion of nonstate actors, larger flexibilities, multiple levels, new steering mechanisms and dispersed loci of authority, among others. These shifts in governance – largely moving away from a monopolistic, nation-state-based, command-and-control environmental governance – resemble ideas of informational governance. But what informational governance adds to the existing insights and frameworks of (new) modes of governance is the centrality of information, informational resources and informational processes. One cannot understand fully the contemporary shifts in environmental governance if informational processes, dynamics, and resources remain unaddressed. The (environmental) governance literature did look into and analyse various developments that also prevail in this book, and we have used their insights on voluntary regulation, information regulation, disclosures, innovative monitoring programs, joint environmental policy making, standards and labels and the like. But only through informational governance glasses one starts to understands the common denominator of these tendencies, and the crucial importance of information and knowledge in contemporary environmental politics. Only then can we grasp the partial relocation of power struggles in environmental politics, understand the restructuring of power relations, assess the full importance of the information revolution for the environment, see how crucial informational resources in environmental struggles are, and recognise the need to analyse networks and (nonmaterial) flows in environmental governance.

Notwithstanding this emphasis and parallel to what the (shifts in) governance literature emphasises, at the moment informational governance is not fully replacing conventional modes, forms, arrangements and resources of environmental governance. Informational governance only partly competes with and replaces conventional state-based institutions and arrangements, whereas to a larger extent it complements and adds to continuing conventional environmental regulation.

Throughout this book, we have seen numerous examples of informational governance, which were backed, facilitated, conditioned or complemented by conventional (national or international) state regulation. In various situations informational governance only could function and exert power when conditioned or legally codified by environmental laws. But, in other situations, informational governance does operate without such state backing.

4. Variations: regions, networks and fluids

In distinguishing and categorising such differences in informational governance, also in relation to conventional modes of governance, the sociology of networks and flows – as presented in Chapter 2 – is helpful. In developing his sociology of networks and flows, Urry (2003) suggests approaching spatial patterns of social relations in three ways or modalities: regions refer to objects geographically clustered together; globally integrated networks consist of more or less stable, enduring and predictable relations between nodes or hubs, stretching across different regions, with relatively walled routes for flows; and global fluids are spatial patterns determined neither by boundaries nor by more or less stable relations, but, by and large, flexibility and liquidity. In each of the three modalities, we can identify different roles for informational governance.

With respect to objects that are geographically clustered together, informational governance is still strongly related to conventional modes of environmental governance, in which states and societies are relevant categories to understand environmental reform. Information disclosures, company environmental reporting, monitoring activities and programs, e-governance and the like are not necessarily performed by state authorities, but often the environmental state does plays a role in coordinating, sanctioning, codifying or facilitating such informational forms of environmental governance. Data-driven regulation (Esty and Rushing, 2006) and informational regulation (Kleindorfer and Orts, 1999) are typical products of informational governance in regions, as they emphasise informational innovations in relation to conventional environmental governance. Moreover, power relations and environmental politics are still strongly tied to conventional resources and certainly not fully replaced to the domain of media, symbolics and information. We also can witness national or regional 'styles' of informational governance, strongly influenced by the national political cultures, policy styles, political and legal systems and the like. Informational governance in China or Japan will look very different from informational governance in the United States or the European Union. Transparency policies, national systems of accountability and legitimacy, public and private systems of verification, the (private or public) structure of the national media, to name but a few, are all contributing to these national variations in informational governance. At the same time, these national specificities are constantly and increasingly challenged by transnational networks that push towards and call for larger harmonisation in a globalising world.

With respect to globally integrated networks, conventional modes of environmental governance are to a larger extent complemented, challenged and sometimes even replaced by new modes of informational governance. With respect to global product and natural resources trade, transnational investment networks, air and train mobility networks and flows, and the like, conventional authoritative resources based on national or international laws, treaties and agreements govern to some extent the environmental side effects. But informational governance arrangements such as global labelling schemes (MSC, FSC, fair trade), companywide environmental reporting of TNCs, environmental information flows through transboundary economic chains and networks, monitoring of and reporting on unsustainable FDI in developing countries, and developments in global sustainable tourism (e.g., van der Duim, 2005) are much less influenced or backed by nation-states. The predictability and structured stability of these network relations make conventional environmental governance modes not impossible, but the transnational character, the mobile flows through these networks and the limited place-boundedness restrict the possibilities of conventional modes of nation-state environmental governance. Informational politics and struggles, changing power relations, questions of reputation and legitimacy and so on are not unique for this modality, but the absence of strong (state-related) legislators, arbiters and/or judges give informational governance more independent room for manoeuvre. Regarding these modalities, informational politics and controversies are less easily settled by nation-states or passed by conventional resources.

Finally, global fluids miss the structured, stable and predictable relations that characterise globally integrated networks and are thus difficult to catch and governed through conventional modes only. Complexity scholars especially focus on this modality, emphasising the uncontrollable, unpredictable, nonlinear and unmanageable fluids when governance is absent. Energy systems and their relations to climate change, the global car system and global fisheries are typical examples of such global fluids that lack a centre, are unstructured and complex and will not easily be governed; at least not through conventional (national or international) governance modes. Here, informational governance can be expected to become dominant, informational resources become powerful and prevailing in environmental governance, and the centre of environmental politics is relocated from state politics to the media, symbols, reputation and legitimation, and non-state global actor networks. Especially when new, transnational and globalised, complex and unstructured, mobile and fluid sustainability problems emerge informational governance seemed to have become the dominant mode of environmental governance.

5. Assessing informational governance: environment and democracy

Finally, the question is, of course, how do we assess informational governance? Although often the initial focus, such an assessment should not only – or not even primarily – concentrate on conventional criteria of environmental effectiveness, as I will argue later in this section. And how should actors in defence of environmental quality strategically operate under conditions of an emerging informational order and governance? Where lie the crucial sites of power and influence for environmentalists (which are no longer to be defined only within the NGO community as ecological modernisation scholars have argued so forcefully), and where are the potential dangers and threats?

In their continuous discussions with the economic and political powers environmentalists in public, private and civil society organisations and arrangements have always been primarily worried about environmental effects of dominant economic patterns, practices and institutions and the environmental effectiveness of various governance institutions and arrangements that address these environmental deteriorations. No matter how wide the environmental agenda has developed – for example, with the introduction of notions of sustainability and quality of life – environmental quality remains at its center.

In assessing new modes of environmental governance a key concern will indeed be related to what might instrumentally be labelled 'environmental effectiveness'. Various studies have indeed concentrated on the environmental effects of ICT systems and of new production and consumption patterns and arrangements that followed the Information Age (see those mentioned in Chapter 1). Others also have tried to assess the effectiveness of new governance approaches and modes that are stronger based on information such as data-driven environmental regulation (cf. Esty, 2006) or voluntary, collaborative and information-based approaches in environmental policy (cf. Norberg-Bohm and de Bruijn, 2005). To be sure and short, it is not possible to give any general and overall conclusion on how effective informational governance protects environmental quality or governs environmental flows. After the categorisation of the variety of informational governance arrangements in three fundamentally different baskets (see earlier), this should not surprise us too much. The environmental success of informational governance can at best be assessed in concrete time-space contexts, where specific arrangements operate in managing specific environmental flows and capital. Although even then results and assessments prove to be difficult to interpret and generalise for various reasons, we should at least be able to formulate conclusions and recommendations on specific improvements. Assessing the environmental successes of concrete (that is, time-space specific) ecolabelling schemes, monitoring and disclosure practices, auditing and verification institutions, and informational campaigns can provide insight on how to improve such schemes and arrangements, whether and how they should be connected with conventional regulatory regimes, regarding what sociomaterial infrastructures and networks they seem to work better and where the sites of power are located to make a difference. But that means at the same time that any general and overall discussion or statement on the value (in terms of their contribution to environmental protection) of informational governance vis-à-vis conventional governance is pointless. In line with some of the scholars working on shifts in governance and (new) modes of governance, we also have to concluded that it is not possible to provide any overall preference for informational versus conventional modes of environmental governance, or vice versa.

But there are other criteria to assess informational governance. A significant part of the environmental movement has always understood themselves as a social movement with a broader agenda. Democracy, equity, and equality, whether it was in the form of participation, equal rights, transparency and/or accountability, often have been related to or even part of the agenda of environmental advocacy groups, although the automatic linkages and relations between a 'green' and a 'red/progressive' agenda have disappeared in the 1980s. Within the conventional model of environmental protection democracy and equality were always strongly related to the state, with different theories and interpretations on the emancipatory, neutral, or capitalist properties of the 'environmental state'. A more democratic and egalitarian environmental governance mode was to be developed through (pressuring) the state. With informational governance the locus of power, democracy and equality is no longer bound up with the state, and environmentalists have understood that quite well and timely. But does informational governance provide for modes and forms of environmental governance that are inherently more democratic and more egalitarian? No. Informational governance is neither inherently more, nor less, democratic or egalitarian. Chadwick and May (2003) are perhaps most clear on this in their assessment of the changing modes of governance in the age of the Internet. Although in principle or in theory the Internet does provide possibilities for a more democratic mode of (e-)governance, they show in detail how in fact three models of e-governance are emerging and practiced (a managerial, a consultative and a participatory) with highly different scores on democracy. This is in line with other research on e-governance, all pointing at conclusions that I would summarise as follows: under informational governance the rules of the game, the institutional layout, the resources that matter – and, with all that, power balances – are changing, but without any inherent conclusion with respect to more or less democracy, equity and equality. The initial sites, origins and sources of power are relocating, and state and non-state environmental advocacies might profit from that. We have extensively analysed throughout this volume how different actors strategically manoeuvre to obtain favourable positions in new power games, also with respect to the environment. Although there is a relation with the old sources of authoritative and economic power, as Paehlke (2003) has shown so strongly in his analysis of the democratic and environmental dangers of 'electronic capitalism', informational power has its own dynamics and resources. So, although informational governance does not turn out to be inherently more democratic and egalitarian, the lines between inclusion and exclusion, or powerful and powerless, are drawn in different ways compared to conventional environmental governance, giving new opportunities and threats for different actors in environmental politics.

This is not fundamentally different at a global level. Developing countries, such as China and Vietnam, have been part of global informational developments to a different extent and in different ways. Although Vietnam, and numerous sub-Saharan African countries, are having a difficult time linking to the key informational highways that connect the most developed and dynamic hubs and nodes of the globe, this is to a lesser extent so for parts of eastern China. This also reflects on new modes of environmental governance. On all informational indicators that are of relevance for informational governance of the environment China as a whole – and even more the Western parts of China – score(s) higher than Vietnam: PCs, Internet connections, environmental movements, environmental labels and standards, media openness and so on. Although it seems justified to conclude that the conditions for informational innovations in environmental governance in China are overall better than in Vietnam (cf. Mol and van Buuren, 2003; Carter and Mol, 2006; see also Chapter 10), it is not easy to draw any conclusions on the better or more effective environmental governance of China vis-à-vis Vietnam (if a comparison between such different countries could be made at all). What could at least be hypothesised is that the further integration of China in the global network economy and society relates to its swifter transformation in environmental governance, towards the inclusion of informational elements. Of course, we should be careful of too simple convergence and homogenisation schemes in global environmental governance, as place-bound local specifics do codetermine modes of environmental governance.[3] But a clear, although complicated, relation exists between the (spatial) patterns of informational governance of the environment and the spreading of new networked forms of the informational economy and society. That means that informational governance is especially adequate for governing institutions, networks and flows that make up this

---

[3] Pollitt (2001) makes a useful classification in public management convergence by distinguishing four forms (and, according to him, stages): discursive convergence, decisional convergence, practice convergence and results convergence. With respect to our assessment and comparison, we can at best expect to find the first two forms of informational governance convergence, and, incidentally, perhaps the third.

new modernity; by the same token, criticising a purely conventional, state-centric model of environmental governance loses adequacy when directed at places, projects and flows marginally related to the 'space of flows', such as in many instances in Vietnam. In contemporary Vietnam, we should expect much more environmental effectiveness from enhancing the capacity of conventional state-led environmental institutions, than from launching new informational strategies.

Hence, there is neither reason for fully embracing informational governance as the last and most innovative sprout of successful environmental reform, nor for condemning it as a toothless, voluntary and deregulatory enterprise because it refrains from state sanctioning power to make a difference. The best assessment is perhaps this: informational governance is emerging, be it in distinct forms and outlooks, in different degrees and speed, at different places, networks and fluids. It will not likely fully replace conventional forms of environmental governance, but, rather, constructs mixes and 'cohabitation' with it, which vary between regions, networks and fluids. Because of this, and because we are most likely only at the beginning of informational governance, there is ample room to further design and set priorities of some forms and arrangements of informational governance over others. What will e-governance look like: managerial, consultative or participatory (Chadwick and May, 2003); what kind of ecolabelling schemes will become dominant (e.g., public, private, mixed); along which lines will inclusions and exclusion of groups from informational flows run in the future; to what extent do we allow homeland security arguments to influence environmental disclosure programs; how will we deal with informational uncertainties in new governance arrangements; what arrangements for verification, auditing and accountability will be designed and implemented; and who will be 'in charge' of these? It is the answers to these and other questions that will in the end determine how we judge informational governance on the environment.

6. New governance modes, new research agendas

Every new development, approach or mode of environmental protection brings along new challenges, new questions and new conflicts. Intensive social and policy science research over the past twenty-five years has made us aware of the shortcomings, limitations and strengths of conventional regulatory regimes of environmental protection. And these limitations are only becoming more evident in an era marked by globalisation and reflexive modernity, resulting in the paradigmatic shift from environmental regulation to environmental governance in the 1990s. The emergence of a new informational mode of environmental governance should be understood as part of this paradigmatic shift. Informational governance, strongly triggered by and fitting into wider developments in late modern society, searches for answers to the weaknesses in conventional environmental regulatory programs. But informational governance also has its own challenges, complications and problems. In setting a new research agenda on informational governance, I will outline four major themes.

The first theme for a research agenda relates to the dynamics and mechanisms of informational governance in various empirical fields, practices and system of environmental protection. Too little is known about the forms, spreading, applications, and modes, and the effects of informational governance of the environment. A more systematic analysis of specific practices would provide us a better idea of its relevance, impact and relation to conventional governance systems. Our provisional categorisation in regions, integrated networks and global fluids could be understood as a first attempt that is in need of more substantial flesh and further detailing.

The second set of questions and challenges relates to structural uncertainty, multiple knowledges and informational overflow. When information and knowledge become crucial resources in and arenas of environmental governance, how do we deal with a constant questioning and revision of environmental knowledge and information, the related uncertainties that seem to form a structural property of environmental reform, and the problem of information overflow, as no undisputed authorities are able to qualify these flows? What kind of new (science-policy) arrangements, decision-making structures and practices, guiding heuristics and principles, and 'closure mechanisms' are, can be, and should be developed in informational environmental governance, in order to prevent stalemate positions? The emergence of informational governance should not make us blind to these questions and challenges, as much as ideas of radical uncertainty, multiple knowledges and information overflow should not overlook (or even condemn) the progressive, transformative powers of environmental information in environmental reform.

A third set of questions is related to new power constellations in informational governance. Inequalities and monopolies related to the construction of environmental knowledge, information handling capacity, information generation and transmission capabilities, access to information and to information publication and credibility construction are becoming key resources in power struggles concerning informational governance. We know that environmental authorities, polluting producers or media multinationals are able to monopolise environmental definitions, environmental monitoring and access to and distribution of environmental information and knowledge, but to what extent does that happen and with what kind of consequences on environmental struggles and outcomes? Where, to what extent, and how do citizens, consumers, insurance companies, environmental NGOs and others gain a position in the 'information battlefield', strengthening their power in new arrangements of informational reform? What kind of sociomaterial infrastructures, in what kind of settings, favour democratic information flows? These questions open up new debates on and challenges for democratic environmental governance, surveillance and countersurveillance, and inclusion and exclusion, on all levels (local to global) and spatial relations (regions, networks, fluids) and without any predetermined outcome.

A final set of (normative) questions emerge with respect to the form and design of informational reform. Some theorists (cf. Young, 1994; Urry, 2003) claim that under conditions of the Information Age the (environmental) state loses its relevance, either in reality or as a concept. At the same time, the more empirical literature on informational *regulation* – in contrast – tries to draw lessons with respect to the concrete design of informational regulation and information disclosure arrangements, especially related to effectiveness, efficiency and costs (e.g., Tietenberg, 1998) and to performance, access and democracy (e.g. Case, 2001; Cohen, 2000). If our social science notion of informational governance can be interpreted as walking between these two, a number of questions become relevant. What material and social infrastructures facilitate and contribute to effective and democratic informational reform of the environment? What requirements does this pose and what possibilities does this open for environmental governance arrangements and networks; for monitoring, transparency and verification structures and strategies; for information handling capacities and capabilities; and for stimulating (information) access rather than

References

Aelst, P. van, and S. Walgrave (2004), "New media, new movements? The role of the internet in shaping the 'anti-globalization' movement". In W. van de Donk, B. D. Loader, P. G. Nixon, and D. Rucht (Eds.), *Cyberprotest: New Media, Citizens and Social Movements*, London: Routledge, pp. 97–122

Ajzen, I., and M. Fishbein (1975), *Belief, Attitude, Intention and Behavior: An Introduction to Theory and Research*, Reading, Mass.: Addison-Wesley

Ajzen, I., and M. Fishbein (1977), "Attitude-behavior relations: A theoretical analysis and review of empirical research", *Psychological Bulletin*, 84, 888–918

Allenby et al. (2001), "Overview and perspectives". In D. J. Richards, B. R. Allenby, and W. D. Compton (Eds.), *Information Systems and the Environment*, Washington, D.C.: National Academy Press

Andersen, M. S. (1994), *Governance by Green Taxes, Making Pollution Prevention Pay*, Manchester: Manchester University Press

Aufderheide, P. (1998), "Niche-market culture, off- and online". In D. L. Borden and K. Harvey (Eds.), *The Electronic Grapevine: Rumour, Reputation and Reporting in the New Online Environment*, London: Lawrence Erlbaum Associates, pp. 43–57

Axelsson, B., and G. Easton (Eds.) (1988), *Industrial Networks: A New View of Reality*, London/New York: Routledge

Ayres, R. U. and U. E. Simonis (eds) (1995), *Industrial metabolism: restructuring for sustainable development*, Tokyo: United Nations University

Badham, R. (1984), "The sociology of industrial and post-industrial societies", *Current Sociology*, 32(1), 1–141

Badham, R. J. (1986), *Theories of Industrial Society*, London/Sydney: Croom Helm

Baker, J. C., B. E. Lachman, D. R. Frelinger, K. M. O'Connell, A. C. Hou, M. S. Tseng, D. Orletsky and C. Yost (2004), *Mapping the Risks: Assessing the Homeland Security Implications of Publicly Available Geospatial Information*, Santa Monica, Calif.: Rand Corporation

Bardoel, J. (1996), "Beyond journalism", *European Journal of Communication*, 1(3), 283–302

Barnett, G. A. (2004), *The Structure of International Internet Hyperlinks and Bilateral Bandwidth*, New York: University of Buffalo, Department of Communication

Barnett, J. (2001), *The Meaning of Environmental Security. Ecological Politics and Policy in the New Security Era*, London/New York: Zed Books

Battistella, G. (with Reporters without Borders) (2005), *Xinhua: The World's Biggest Propaganda Agency*, Paris: Reporters without Borders

Baudrillard, J. (1981), *For a Critique of the Political Economy of the Sign*, St. Louis: Telos Press (translation of *Pour une Critique de l'économie politique du signe*. Collection Les Essais, 168. Gallimard, Paris, 1972)

Bauman, Z. (1987), *Legislators and Interpreters*, Cambridge: Polity

Bauman, Z. (1992), *Intimations of Postmodernity*, London: Routledge

Bauman, Z. (2000), *Liquid Modernity*, Cambridge: Polity

Beck, U. (1986), *Risikogesellschaft: Auf dem Weg in eine andere Moderne* (Risk Society: On the Way toward Another Modernity). Frankfurt am Main: Suhrkamp

Beck, U. (1994), "The reinvention of politics: Towards a theory of reflexive modernisation". In U. Beck, A. Giddens, and S. Lash (Eds.), *Reflexive Modernisation: Politics, Tradition and Aesthetics in the Modern Social Order*, Cambridge, U.K.: Polity, pp. 1–55

Beck, U. (1996), "World risk society as cosmopolitan society? Ecological questions in a framework of manufactured uncertainties", *Theory, Culture & Society*, 13(4), 1–32

Beck, U. (1997), *Was ist Globalisierung? Irrtümer des Globalismus – Antworten auf Globalisierung*, Frankfurt am Main: Suhrkamp

Beck, U. (2004), *Power in the Global Age*, London: Polity

Beck, U., A. Giddens and S. Lash (1994), *Reflexive Modernization: Politics, Tradition and Aesthetics in the Modern Social Order*, Cambridge: Polity Press

Beck, U., and J. Willms (2004), *Conversations with Ulrich Beck*, Cambridge: Polity

Beierle, T. C. (2004), "The benefits and costs of disclosing information about risks: What do we know about right-to-know?", *Risk Analysis*, 24(2), 335–346

Bell, D. (1973), *The Coming of Post-industrial Society: A Venture in Social Forecasting*, Harmondsworth: Penguin

Bell, D. (1976), *The Cultural Contradictions of Capitalism*, London: Heinemann

Bell, D. (1979), "The social framework of the information society". In M. L. Dertouzous and J. Moses (Eds.), *The Computer Age: A Twenty-Year View*, Cambridge, Mass.: MIT, pp. 163–21

Bell, D. (1980) "The Social Framework of the Information Society", in T. Forester (ed.) *The Microelectronics Revolution*, Oxford: Blackwell, pp. 500–549

Bell, D. (1987), "The world and the United States in 2013", *Daedalus*, 116, 1–31

Bell, M. M. (2004), *An Invitation to Environmental Sociology*, 2nd ed., Thousands Oaks, Calif.: Pine Forge Press

Beniger, J. R. (1986), *The Control Revolution: Technological and Economic Origins of the Information Society*, Cambridge, Mass.: Harvard University Press

Bennett, W. L. (2004), "Communicating global activism: Some strength and vulnerabilities of networked politics". In W. van de Donk, B. D. Loader, P. G. Nixon and D. Rucht (Eds.), *Cyberprotest: New Media, Citizens and Social Movements*, London: Routledge, pp. 123–146

Bennie, L. (1998), "Brent Spar, Atlantic Oil and Greenpeace". In F. Ridley and G. Jordan (Eds.), *Protest Politics: Cause and Campaigns*, Oxford/New York: Oxford University Press

Berg, C. (2003), *Rebound Effects – Their Nature and Their Challenge*, Paper presented in the TERRA 2000 project (http://www.tukkk.fi/tutu/terra2000/)

Biggart, N. W., and T. D. Beamish (2003), "The economic sociology of conventions: Habit, custom, practice and routine in market order", *Annual Review of Sociology*, 29, 443–464

Bloomfield, D., K. Collins, C. Fry and R. Munton (2001), "Deliberation and inclusion: Vehicles for increasing trust in UK public governance?" *Environment and Planning C: Government and Policy*, 19, 501–513

Blowers, A. (1997), "Environmental policy: Ecological modernization and the risk society?", *Urban Studies*, 34(5–6), 845–871

Blühdorn, I. (2000), "Ecological modernisation and post-ecologist politics". In G. Spaargaren, A. P. J. Mol, and F. Buttel (Eds.), *Environment and Global Modernity*, London: Sage, pp. 209–228

Blumer, J., and M. Gurevitch (1995), *The Crisis of Public Communication*, London: Routledge

Boltanski, L., and L. Thévenot (1991), *De la justification: Les économies de la grandeur*, Paris: Gallimard

Bookchin, M. (1982), *The Ecology of Freedom: The Emergence and Dissolution of Hierarchy*, Palo Alto, Calif.: Cheshire Books

Boyd-Barrett, O. (2004), "U.S. global cyberspace". In D. Schuler and P. Day (Eds.), *Shaping the Network Society: The New Role of Civil Society in Cyberspace*, Cambridge, Mass.: MIT, pp. 19–42

Braman, S. (2006), *Change of State: Information, Policy, and Power*, Cambridge, Mass.: MIT

Braverman, H. (1974), *Labour and Monopoly Capital: The Degradation of Work in the Twentieth Century*, New York and London: Monthly Review Press

Brettell, A. (2004), *Nipping Dissent in the Bud: The Institutionalization of Environmental Complaint Resolution in China*, paper presented at the conference Shifting Social Spaces, April 22–23, 2004, Tsinghua University, Beijing

Brown, L. R. (2001), *Eco-economy: Building an Economy for the Earth*, New York: W. W. Norton

Bruijn, T. de, and V. Norberg-Bohm (Eds.) (2005), *Industrial Transformation: Environmental Policy Innovation in the United States and Europe*, Cambridge, Mass.: MIT

Bulkeley, H., and A. P. J. Mol (2003), "Participation and environmental governance: Consensus, ambivalence and debate", *Environmental Values*, 12, 143–154

Burg, S. W. K. van den (2004), "Informing or empowering? Disclosure in the United States and the Netherlands", *Local Environment: The International Journal of Justice and Sustainability*, 9(4), 367–382

Burg, S. W. K. van den (2006), *Governance through Information: Environmental Monitoring from a Citizen-Consumer Perspective*, Wageningen: Wageningen University (dissertation environmental policy)

Burg, S. W. K. van den, A. P. J. Mol and G. Spaargaren (2003), "Consumer-oriented monitoring and environmental reform", *Environment and Planning C: Government and Policy*, 21(3), 371–388

Burnham, D. (1983), *The Rise of the Computer State*, London: Weidenfeld and Nicholson

Burström, F., and A. Lindqvist (2002), "Environmental information management in municipalities", *Local Environment*, 7(2), 189–201

Buttel, F. H. (1978), "Environmental sociology: A new paradigm?", *American Sociologist*, 13, 252–256

Buttel, F. H. (2000), "Ecological modernization as social theory", *Geoforum*, 31(1), 57–65

Buttel, F. H. (2002), "Has environmental sociology arrived?", *Organization & Environment*, 15(1), 42–55

Buttel, F. H. (2003), "Environmental sociology and the explanation of environmental reform", *Organization & Environment*, 16(3), 306–344

Buttel, F. H. (2006), "Globalization, environmental reform and U.S. hegemony". In G. Spaargaren, A. P. J. Mol, and F. H. Buttel (Eds.), *Governing Environmental Flows: Global Challenges for Social Theory*, Cambridge, Mass.: MIT, pp. 157–184

Buttel, F. H., A. P. Hawkins, and A. G. Power (1992), "From limits to growth to global change: Constraints and contradictions in the evolution of environmental science and ideology", *Global Environmental Change*, December, pp. 57–66

Callon, M., C. Méadel and V. Rabeharisoa (2002), "The economy of qualities", *Economy and Society*, 31(2), 194–217

Campbell, D. J. (2003), "Intra- and intersectoral effects in environmental disclosures: Evidence for legitimacy theory?", *Business Strategy and the Environment*, 12, 357–371

Carolan, M. S., and M. M. Bell (2003), "In truth we trust: Discourse, phenomenology, and the social relations of knowledge in an environmental dispute", *Environmental Values*, 12, 225–245

Carolan, M. (2004), "Ecological Modernization: What about Consumption?", *Society and Natural Resources* 17, 3, pp. 247–260

Carson, R. (1962), *Silent Spring*, Middlesex (UK): Penguin

Carter, N. (2000), *The Politics of the Environment: Ideas, Activism, Policy*, Cambridge: Cambridge University Press

Carter, N. T., and A. P. J. Mol (2006), "China and environment: Domestic and transnational dynamics of a future hegemon", *Environmental Politics*, 15(2), 331–345

Case, D. W. (2001), "The law and economics of environmental information as regulation", *Environmental Law Reporter*, 31, 10773–10789

Castells, M. (1996), *The Rise of the Network Society: Volume I of The Information Age: Economy, Society and Culture*, Malden, Mass./Oxford: Blackwell

Castells, M. (1997a), *The Power of Identity: Volume II of The Information Age: Economy, Society and Culture*, Malden, Mass./Oxford: Blackwell

Castells, M. (1997b), *End of Millennium: Volume III of The Information Age: Economy, Society and Culture*, Malden, Mass./Oxford: Blackwell

Castells, M. (2001), *The Internet Galaxy: Reflections on the Internet, Business, and Society*, Oxford: Oxford University Press

Castells, M. (2004), "Informationalism, networks, and the network society: A theoretical blueprint". In M. Castells (Ed.), *The Network Society: A Cross-Cultural Perspective*, Cheltenham: Edward Elgar, pp. 3–45

Catton, W. R., and R. E. Dunlap (1978a), "Environmental sociology: A new paradigm", *American Sociologist*, 13, 41–49

Catton, W. R., and R. E. Dunlap (1978b), "Paradigms, theories, and the primacy of the HEP-NEP distinction", *American Sociologist*, 13, 256–259

CCW Research (2002), *The Truth about Government Websites in China* (in Chinese). (http//www.ccwresearch.com.cn/cn/news/0912_01.asp)

Chadwick, A., and C. May (2003), "Interaction between states and citizens in the age of the Internet: 'E-government' in the United States, Britain, and the European Union", *Governance: An International Journal of Policy, Administration, and Institutions*, 16(2), 271–300

Champbell, D., and S. Connor (1986), *On the Record: Surveillance, Computers and Privacy*, London: Michael Joseph

Chandler, W., R. Schaeffer and Z. Dadi et al. (2002), *Climate Change Mitigation in Developing Countries: Brazil, China, India, Mexico, South Africa, and Turkey*, Arlington, Va.: Pew Center on Global Climate Change

Chang, S., H.-M. Chiu and W.-L. Tu (2006), "Breaking the silicon silence: Voicing health and environmental impacts within Taiwan's Hsinchu Science Park". In T. Smith, D. A. Sonnenfeld and D. N. Pellow (Eds.), *Challenging the Chip: Labor Rights and Environmental Justice in the Global Electronics Industry*, Philadelphia: Temple University Press, pp. 170–180

Chapman, G. (2004), "Shaping technology for the 'Good Life': The technological imperative versus the social imperative". In D. Schuler and P. Day (Eds.), *Shaping the Network Society: The New Role of Civil Society in Cyberspace*, Cambridge, Mass.: MIT, pp. 43–66

Chesters, G., and I. Welsh (2005), "Complexity and social movement(s): Process and emergence in planetary action systems", *Theory, Culture & Society*, 22(5), 187–211

Christoff, P. (1996), "Ecological modernisation, ecological modernities", *Environmental Politics*, 5(3), 476–500

Clapp, J. (2005), "The privatization of global environmental governance: ISO 14000 and the developing world". In D. L. Levy and P. J. Newell (Eds.), *The Business of Global Environmental Governance*, Cambridge, Mass.: MIT, pp. 223–248

Cleaver, H. (1998), "The Zapatistas and the electronic fabric of struggle". In J. Holloway and E. Pelaez (Eds.), *Zapatista! Reinventing Revolution in Mexico*, London: Pluto Press

Coglianese, C. (2004), *E-Rulemaking; Information Technology and Regulatory Policy*, Regulatory Policy Report No. RPP-05 (http://snipurl.com/82pi)

Cohen, M. A. (1997), "Risk society and ecological modernisation: Alternative visions for post-industrial nations", *Futures*, 29(2), 105–119

Cohen, M. A. (2000), *Information as a Policy Instrument in Protecting the Environment: What Have We Learned?* (Environmental Defense Fund, Washington, D.C.) (http://www.environmentaldefense.org/wip/cohen.html)

Cohen, M. A. (2002), "Transparency after 9/11: Balancing the 'right-to-know' with the need for security", *Corporate Environmental Strategy*, 9, 368–374

Connelly, J., and G. Smith (2003), *Politics and the Environment: From Theory to Practice*, 2nd ed., London: Routledge

Cottle, S. (2000), "Rethinking theories of news access", *Journalism Studies*, 1(3), 427–448

Cranor, C. (2005), "Precautionary information-generation in science and the law", *Water Science & Technology*, 52(6), 65–71

Cushing, K. K., H. McGray and H. Lu (2005), "Understanding ISO 14001 adoption and implementation in China", *International Journal of Environment and Sustainable Development*, 4(3), 246–268

Dalton, R. J. (2005), "The greening of the globe? Cross-national levels of environmental group membership", *Environmental Politics*, 14(4), 441–459

Dandeker, C. (1990), *Surveillance, Power and Modernity: Bureaucracy and Discipline from 1700 to the Present Day*. Cambridge, U.K.: Polity Press

Dasgupta, S., J. H. Hong, B. Laplante and N. Mamingi (2004), *Disclosure of Environmental Violations and the Stock Market in the Republic of Korea*, Policy Research Working Paper 3344, Washington, D.C.: World Bank

Dasgupta, S., and D. Wheeler (1996), *Citizen Complaints as Environmental Indicators: Evidence from China*, World Bank Policy Research Working Paper, Washington, D.C.: World Bank

Davidson, D. J., and S. Frickel (2004), "Understanding environmental governance: A critical review", *Organization & Environment*, 17(4), 471–492

Davies, A. (2001), "What silence knows – planning, public participation and environmental values", *Environmental Values*, 10, 77–102

De Burgh, H. (2003), *The Chinese Journalist*, London: Routledge

De Marchi, B., S. Functowicz, C. Gough, et al. (1998), *The ULYSSES Voyage, the ULYSSES Project at the JRC*, Ispra, Joint Research Centre, European Commission EUR 17760EN

Deegan, C., and M. Rankin (1996), "Do Australian companies report environmental news objectively?" *Accounting, Auditing and Accountability Journal*, 9(2), 50–67

della Porta, D., and L. Mosca (2005), "Global-net for global movements? A network of networks for a movement of movements?", *Journal of Public Policy*, 25(1), 165–190

DeSombre, E. R. (2000), *Domestic Sources of International Environmental Policy: Industry, Environmentalists, and U.S. Power*, Cambridge, Mass.: MIT

Di Chang-Xing (1999), "ISO 14001: The severe challenge for China". In W. Wehrmeyer and Y. Mulugetta (Eds.), *Growing Pains: Environmental Management in Developing Countries*, Sheffield: Greenleaf, pp. 101–116

Diani, M. (2001), "Social movements networks: Virtual and real". In F. Webster (Ed.), *Culture and Politics in the Information Age: A New Politics?*, London: Routledge, pp. 117–128

DiGregorio, M., A. Y. Rambo and M. Yanagisawa (2003), "Clean, green and beautiful: Environment and development under the renovation economy". In Hy. V. Luong (Ed.), *Post War Vietnam: Dynamics of a Transforming Society*, Singapore: Lanham

Dispensa, J. M., and R. J. Brulle (2003), "Media's social construction of environmental issues: Focus on global warming – a comparative study", *International Journal of Sociology and Social Policy*, 23(10), 74–105

Dodge, M., and R. Kitchen (2001), *Mapping Cyberspace*, London: Routledge

Donella H. Meadows, D. H., D. L. Meadows, J. Randers and W. W. Behrens III (1972), *The Limits to Growth*. New York: Universe Books.

Donk, W. van de, B. D. Loader, P. G. Nixon and D. Rucht (Eds.) (2004a), *Cyberprotest: New Media, Citizens and Social Movements*, London: Routledge

Donk, W. van de, B. D. Loader, P. G. Nixon and D. Rucht (2004b), "Introduction: Social movements and ICTs". In W. van de Donk, B. D. Loader, P. G. Nixon and D. Rucht (Eds.), *Cyberprotest: New Media, Citizens and Social Movements*, London: Routledge, pp. 1–25

Downing, J. D. H. (2005), "Activist media, civil society and social movements". In W. de Jong, M. Shaw, and N. Stammers (Eds.), *Global Activism, Global Media*, London/Ann Arbor: Pluto Press, pp. 149–164

Drucker (1969), *The Age of Discontinuity. Guidelines to our Changing Society*, New York: Harper and Row

Dryzek, J. S. (1987), *Rational Ecology: Environment and Political Economy*, Oxford/New York: Blackwell

Dryzek, J. S. (1990), *Discursive Democracy: Politics, Policy and Political Science*, Cambridge: Cambridge University Press

Dryzek, J. S. (1997), *The Politics of the Earth: Environmental Discourses*, Oxford: Oxford University Press

Duim, V. R. van der (2005), *Tourismscapes: An Actor-Network Perspective on Sustainable Tourism Development*, Wageningen: Wageningen University (Ph.D. dissertation)

Dunlap, R. E. (1980), "Paradigmatic change in social science", *American Behavioral Scientist*, 24(1), 5–14

Dunlap, R. E. (2002), "Environmental sociology: A personal perspective on its first quarter century", *Organization and Environment*, 14(1), 10–30

Eckersley, R. (2004), *The Green State: Rethinking Democracy and Sovereignty*. Cambridge, MA.: MIT Press.

Ecologist (1972), *A Blueprint for Survival*, Harmondsworth: Penguin

Economy, E. (2005), *China's Environmental Movement*. Testimony before the Congressional Executive Commission on China Roundtable on Environmental NGOs in China: Encouraging Action and Addressing Public Grievances, Washington, D.C.

Edelman, M. (1964), *The Symbolic Uses of Politics*, Urbana: University of Illinois Press

Edwards, E. (2006), "Empowering communities for environmental decision-making: Innovative partnerships in Cleveland (USA)". In V. J. J. M. Bekkers, H. van Duivenboden and M. Thaens (Eds.), *Information and Communication Technology and Public Innovation: Assessing the ICT-Driven Modernization of Public Administration*, Amsterdam: IOS Press, pp. 175–191

Eisner, M. A. (2004), "Corporate environmentalism, regulatory reform, and industry self-regulation: Toward genuine regulatory reinvention in the United States", *Governance: An International Journal of Policy, Administration, and Institutions*, 17(2), 145–167

Ekins, P., and S. Speck (2000), "Proposals of environmental fiscal reforms and the obstacles to their implementation", *Journal of Environmental Policy and Planning*, 2(2), 93–114

Enzensberger, H. M. (1974 [1973]), "A critique of political ecology", *New Left Review*, 84, 3–32

Erdmann, L., L. Hilty, J. Goodman and P. Arnfalk (2004), *The Future Impact of ICTs on Environmental Sustainability*, Brussels: European Commission (Report No. EUR 21384 EN)

Esty, D. (2001a), "Toward data-driven environmentalism: The Environmental Sustainability Index", *Environmental Law Reporter*, 31(5), 10603–10612

Esty, D. (2001b), "Digital earth: Saving the environment", *OECD Observer*, 226, http://www.oecdobserver.org/news/fullstory.php/aid/627/Digital_earth:_saving_the_environment.html

Esty, D. (2004), "Environmental protection in the information age", *NYU Law Review*, 79(1), 115–211

Esty, D., and R. Rushing (2006), *Data-Driven Policymaking*, Washington, D.C.: Centre of American Progress

European Commission (1992), *Europeans and the Environment*, Eurobarometer 37.0, Brussels: DG XI EC

European Commission (1995), *Europeans and the Environment*, Eurobarometer 43.1 bis, Brussels: DG XI EC

European Commission (1999), *Environnement? Ce que les Européens en pensent*, Eurobarometer 51.1, Brussels: DG XI/DG X EC

European Commission (2002), "The Attitudes of European Citizens towards Environment", *Official Journal of the European Communities* L242/1;

European Commission (2005), *The Attitudes of European Citizens towards Environment*, Eurobarometer No. 217, Brussels: DG Environment EC.

Fang, H. (2002), *The History of Journalism and Communication in China*, Beijing: Renmin University Press

Faucher, I. (2006), *Fighting Industrial Pollution in Viet Nam: Examining the Role and Importance of Citizen Complaints and Public Disclosure Strategies*, Toronto: University of Toronto (Master of Arts thesis, Dept. of Geography)

Fischer-Kowalski, M. (1996), "Operational Paths Towards Sustainability Reconsidered: Changes in Social Metabolism and Colonization of Nature", *Abhandlungen der Geologischen Bundesanstalt* 53, pp. 83–90.

Fishbein, M., and I. Ajzen (1975), *Belief, Attitude, Intention and Behavior*, Reading, Mass.: Addison-Wesley

Florini, A. (2003), *The Coming Democracy: New Rules for Running a New World*, Washington, D.C.: Island Press

Forseback, L. (2000), *The Knowledge Economy and Climate Change: An Overview of New Opportunities*, Brussels: European Commission DG Information Society

Foster, J. B. (1999), "Marx's theory of metabolic rift: Classical foundations for environmental sociology", *American Journal of Sociology*, 105, 366–405

Foster, J. B. (2002). *Ecology against capitalism*. New York: Monthly Review Press.

Foucault, M. (1977), *Discipline and Punish: The Birth of the Prison*, London: Penguin

Frankel, B. (1987), *The Post-industrial Utopians*, Cambridge, U.K.: Polity

Freudenburg, W. (2000), "Social constructions and social constrictions: Toward analyzing the social construction on the 'naturalized' as well as the 'natural'". In G. Spaargaren, A. P. J. Mol, and F. H. Buttel (Eds.), *Environment and Global Modernity*, London: Sage, pp. 103–120

Friends of Nature (2000), *Chinese Newspapers' Environmental Awareness*, Beijing: Friends of Nature

Fung, A., and D. O'Rourke (2000), "Reinventing environmental regulation from the grassroots up: Explaining and expanding the success of the toxics release inventory", *Environmental Management*, 25(2), 115–127

Gaber, I., and A. W. Willson (2005), "Dying for diamonds: The mainstream media and the NGOs – A case study of ActionAid". In W. de Jong, M. Shaw and N. Stammers (Eds.), *Global Activism, Global Media*, London/Ann Arbor: Pluto Press, pp. 95–109

Gao, N. (2005), *Eco-labels on Chinese Cotton Textiles as a Greening Trade Strategy*, Wageningen: Wageningen University (M.Sc. thesis)

Garnham, N. (1998) 'Information Society Theory as Ideology: A Critique', *Loisir et Société* 21(1): 97–120

Gavin, E., and J. Gyamfi-Aidoo (Eds.) (2001), *Environmental Information Systems Development in Sub-Saharan Africa: Approaches, Lessons and Challenges*, Pretoria, South Africa: EIS-Africa

Gereffi, G., and M. Korzeniewicz (Eds.) (1994), *Commodity Chains and Global Capitalism*, Westport, Conn.: Greenwood Press

Gibbs, D. (2004), "Towards an Environmental Economic Geography", unpublished paper, Department of Geography, University of Hull, Hull

Giddens, A. (1976), *New Rules of Sociological Method: a Positive Critique of interpretative Sociologies*. London: Hutchinson

Giddens, A. (1984), *The Constitution of Society*, Cambridge, U.K.: Polity

Giddens, A. (1985), *The Nation-State and Violence*, Cambridge, U.K.: Polity

Giddens, A. (1990), *The Consequences of Modernity*, Cambridge, U.K.: Polity

Giddens, A. (1991), *Modernity and Self-Identity: Self and Society in the Late Modern Age*, Cambridge, U.K.: Polity

Giddens, A. (1994), *Beyond Left and Right: The Future of Radical Politics*, Cambridge, U.K.: Polity

Glasbergen, P., F. Biermann, and A. P. J. Mol (Eds.) (2007), *Partnerships, Governance and Sustainable Development: Reflections on Theory and Practice*, London: Edward Elgar

Glasser, T. L., and S. Craft (1998), "Public journalism and the search for democratic ideals". In T. Liebers and J. Curran (Eds.), *Media Ritual and Identity*, London: Routledge

Glyn, A., and B. Sutcliffe (1992), "'Global but leaderless'? The new capitalist order". In R. Miliband and L. Panitch (Eds.), *New World Order: The Socialist Register*, London: Merlin Press, pp. 76–95

Goldblatt, D. (2002), *Personal vs. Socio-technical Change: Informing and Involving Householders for Sustainable Energy Consumption*, Zurich, Swiss Federal Institute of Technology (http://e-collection.ethbib.ethz.ch) (Ph.D. dissertation)

Goldsmith, J., and T. Wu (2006), *Who Controls the Internet? Illusions of a Borderless World*, Oxford: Oxford University Press

Goldson, A., and J. Murphy (1997), "Ecological modernization: Economic restructuring and the environment", *Political Quarterly*, 68(5), 74–86

Goodman, D., and M. Watts (Eds.) (1997), *Globalising Food: Agrarian Questions and Global Restructuring*, London: Routledge

Gorz, A. (1982), *Farewell to the Working Class: An Essay on Post-industrial Socialism*, London: Pluto Press

Gouldner, A. W. (1979), *The Future of Intellectuals and the Rise of the New Class*, London/New York: Macmillan

Gouldson, A. and J. Murphy (1997), "Ecological modernization: restructuring industrial economies". In: M. Jacobs (ed.), *Greening the millennium? The new politics of the environment*, Oxford: Blackwell, pp. 74–86

Grabher, G. (Ed.) (1993), *The Embedded Firm: On the Socioeconomics of Industrial Networks*, London/New York: Routledge

Graham, M. (2001), *Information as Risk Regulation: Lessons from Experience*, Regulatory Policy Program Working Paper RPP-2001-04, Cambridge, Mass.: Centre for Business and Government, Harvard University

Graham, M. (2002), *Democracy by Disclosure: The Rise of Technopopulism*, Washington, D.C.: Brookings Press

Graham, M., and C. Miller (2005), "Disclosure of toxic releases in the United States". In T. de Bruijn and V. Norberg-Bohm (Eds.), *Industrial Transformation: Environmental Policy Innovation in the United States and Europe*, Cambridge, Mass.: MIT, pp. 307–333

Graham, S., and S. Marvin (2001), *Splintering Urbanism, Networked Infrastructures, Technological Mobilities and the Urban Condition*, London: Routledge

Gramson, W. A. (1995), "Constructing Social Protest". In H. Johnston and B. Klandermans (Eds.), *Social Movements and Culture*, London: UCL Press

Grant, D., and A. W. Jones (2004), "Do manufacturers pollute less under the regulation-through-information regime? What plant-level data tell us", *Sociological Quarterly*, 45(3), 471–486

Grignou, B. le, and C. Patou (2004), "ATTAC(k)ing expertise: Does the Internet really democratize knowledge?" In W. van de Donk, B. D. Loader, P. G. Nixon, and D. Rucht (Eds.), *Cyberprotest: New Media, Citizens and Social Movements*, London: Routledge, pp. 164–179

Gross, M. (2006), *Ignorance and Surprise: Science, Knowledge Production, and the Making of Robust Ecological Design*, Bielefeld: Universität Bielefeld, Habilitationsschrift Facultät Soziologie

Grubler, A., et al. (Eds.) (2002), *Technological Change and the Environment*, Washington, D.C.: Resources For the Future Press

Gunaratne, S. (2002), "An evolving triadic world: A theoretical framework for global communication research", *Journal of World-Systems Research*, 8(3), 330–365

Gunningham, N., and J. Prest (1993), "Environmental audit as a regulatory strategy: Prospects and reform", *Sydney Law Review*, 15(4), 492–526

Guo Peiyuan (2005), *Corporate Environmental Reporting and Disclosure in China*, Hong Kong/Shenzhen: Corporate Social Resposibility Asia

Habermas, J. (1981), *Theorie des kommunikativen Handels*, volumes 1 and 2, Frankfurt: Suhrkamp

Haggerty, K. D., and R. V. Ericson (2000), "The surveillant assemblage", *British Journal of Sociology*, 51(4), pp. 605–622

Hajer, M. A. (1995), *The Politics of Environmental Discourse: Ecological Modernisation and the Policy Process*, Oxford: Clarendon

Håkansson, H., and J. Johanson (1993), "The network as a governance structure: Interfirm cooperation beyond markets and hierarchies". In

G. Grabher (Ed.), *The Embedded Firm: On the Socioeconomics of Industrial Networks*, London/New York: Routledge, pp. 35–51

Ham, C., and M. Hill (1984), *The Policy Process in the Modern Capitalist State*, New York: Harvester Wheatsheaf

Hamelink, C. J. (1995), *World Communication: Disempowerment and Self-Empowerment*, London: Zed Books

Hamelink, C. J. (2004), "Human rights in the global billboard society". In D. Schuler and P. Day (Eds.), *Shaping the Network Society: The New Role of Civil Society in Cyberspace*, Cambridge, Mass.: MIT, pp. 67–81

Hamilton, J. T. (2005), *Regulation through Revelation: The Origin, Politics, and Impacts of the Toxics Release Inventory Program*, Cambridge: Cambridge University Press

Hannigan, J. A. (1995), *Environmental Sociology: A Social Constructionist Perspective*, London and New York: Routledge

Hannigan, J. A. (2006), *Environmental Sociology*, 2nd ed., London and New York: Routledge

Hansen, A. (Ed.) (1993a), *The Mass Media and Environmental Issues*, Leicester: Leicester University Press

Hansen, A. (1993b), "Greenpeace and press coverage of environmental issues". In A. Hansen (Ed.), *The Mass Media and Environmental Issues*, Leicester: Leicester University Press, pp. 150–178

Hansen, F. (2000), "Three emerging issues, three needed changes", *Environmental Forum*, May/June

Hansjürgens, B. (2000), "Symbolische umweltpolitik – eine erklärung aus sicht der neuen politischen ökonomie". In B. Hansjürgens and G. Lübbe-Wolf (Eds.) (2000), *Symbolische Umweltpolitik*, Frankfurt am Main: Suhrkamp, pp. 144–182

Hansjürgens, B., and G. Lübbe-Wolf (Eds.) (2000), *Symbolische Umweltpolitik*, Frankfurt am Main: Suhrkamp

Harland, P. (2001), *Pro-environmental Behavior*, Leiden: University of Leiden (dissertation)

Harwitt, E. (2004), "Spreading telecommunications to developing areas in China: Telephones, the Internet and the digital divide", *China Quarterly*, 180, 1010–1030

Heinonen, S., P. Jokinen, and J. Kaivo-oja (2001), "The ecological transparency of the information society", *Futures*, 33, 319–337

Held, D., and M. Koenig-Archibugi (Eds.) (2003), *Taming Globalization: Frontiers of Governance*, Cambridge, U.K.: Polity

Henson, S., and S. Jaffee (2007), "Developing country responses to the enhancement of food safety standards". In U. Grote, A. Basu, and N. Chau (Eds.), *New Frontiers in Environmental and Social Labelling*, Heidelberg: Physica-Verlag/Springer, pp. 193–220

Héritier, A. (2002), "New modes of governance in Europe: Policy-making without legislating?" In A. Heritier (Ed.), *Common Goods: Reinventing European and International Governance,* Lanham, Md.: Rowman and Littlefield

Herrick, C. N. (2004), "Objectivity versus narrative coherence: Science, environmental policy, and the U.S. Data Quality Act", *Environmental Science & Policy,* 7, 419–433

Hills, P., and C. S. Man (1998), "Environmental regulation and the industrial sector in China: The role of informal relationships in policy implementation", *Business Strategy and the Environment,* 7, 53–70

Hinchliffe, S. (2001), "Indeterminacy in-decisions – science, policy and politics in the BSE (Bovine Spongiform Encephalopathy) crisis", *Transactions of the Institute of British Geographers,* 26, 182–204

Hirsch, D. D. (2001), "Globalization, information technology, and environmental regulation: An initial inquiry", *Virginia Environmental Law Journal,* 20(1), 57–74

Hirst, P., and G. Thompson (1996), *Globalisation in Question?,* London: Polity

Ho, P. (2001), "Greening without conflict? Environmentalism, NGOs and civil society in China", *Development and Change,* 32(5), 893–921

Holgaard, J. E., and T. H. Jørgensen (2005), "A decade of mandatory environmental reporting in Denmark", *European Environment,* 15, 262–273

Holtz, S., and T. Demopoulos (2006), *Blogging for Business: Everything You Need to Know and Why You Should Care,* New York: Kaplan Business

Holz, C. A. (2003), "'Fast, clear and accurate': How reliable are Chinese output and economic growth statistics?", *China Quarterly,* 173, 122–163

Hong Lui (1998), "Profit or ideology? The Chinese press between party and market", *Media, Culture & Society,* 20, 31–41

Hong, J. H., B. Laplante, and C. Meisner (2003), *Public Disclosure of Environmental Violations in the Republic of Korea,* World Bank Policy Research Working Paper 3126, Washington, D.C.: World Bank

Hordijk, L. et al. (2002), *Climate Options for the Long Term: Final Report,* Wageningen: Wageningen University

Horton, D. (2004), "Local environmentalism and the Internet", *Environmental Politics,* 13(4), 734–753

Horvath, S. (2002), "New law will let business attack data underlying rules", *Wall Street Journal,* July 5, p. 25

Howes, M. (2001), *Digital Disclosures: Environmental Governance and the Internet,* paper presented at the conference "New Natures, New Cultures, New Societies", Fitzwilliam College, Cambridge, 5–7 July 2001

Howes, M. (2002), "Reflexive modernization, the internet, and democratic environmental decision making", *Organization & Environment*, 15(3), 328–331

Huber, J. (1991), *Unternehmen Umwelt. Weichenstellungen für eine ökologische Marktwirtschaft*, Frankfurt am Main: Fisher

Hu, Kanping (2001), "Harmony in diversity: The relationship between environmental journalism and green NGOs in China". In J. Turner and F. Wu (Eds.), *Green NGOs and Environmental Journalist Forum: A Meeting of Environmentalists in Mainland China, Hong Kong and Taiwan*, Washington, D.C.: The Woodrow Wilson Centre, pp. 30–31

Hu, Kanping, and Xiaogang Yu (2005), "Bridge over troubled waters: The role of the news media in promoting public participation in river basin management and environmental protection in China". In J. L. Turner and K. Otsuka (Eds.), *Promoting Sustainable River Basin Governance: Crafting Japan-U.S. Water Partnerships in China*, Chiba: Institute of Developing Economies (IDE Spot Survey no. 28), pp. 125–140

Huber, J. (1982), *Die verlorene Unschuld der Ökologie: Neue Technologien und superindustrielle Entwicklung*, Frankfurt am Main: Fisher Verlag

Huber, J. (2004), *New Technologies and Environmental Innovation*, Cheltenham: Edward Elgar

Humphrey, J., and H. Schmitz (2002), *Developing Country Firms in the World Economy: Governance and Upgrading in Global Value Chains*, INEF report 61/2002, Duisburg: University of Duisburg (INEF)

Inglehart, R. (1977), *The Silent Revolution: Changing Values and Political Styles among Western Publics*, Princeton, N.J.: Princeton University Press

Inglehart, R. (1990), *Culture Shift in Advanced Industrial Society*, Princeton, N.J.: Princeton University Press

Irwin, A. (1995), *Citizen Science: A Study of People, Expertise and Sustainable Development*, London and New York: Routledge

Isenmann, R., and C. Lenz (2002), "Internet use for corporate environmental reporting: Current challenges – technical benefits – practical guidance", *Business Strategy and the Environment*, 11, 181–202

Jacek, H. J. (1991), "The functions of associations as agents of public policy". In A. Martinelli (Ed.), *International Markets and Global Firm: A Comparative Study of Organized Business in the Chemical Industry*, London: Sage

Jaffer, R., T. M. Iskandar and J. N. Muhamad (2002), "An investigation of environmental disclosures: Evidence from selected industries in Malaysia", *International Journal of Business and Society*, 3(2), 55–68

Jänicke, M. (1986), *Staatversagen: Die Ohnmacht der Politik in die Industriegesellschaft*, Munich: Piper (translated as *State Failure: The Impotence of Politics in Industrial Society*, Cambridge, U.K.: Polity Press, 1990)

Jänicke, M. (2006), "The environmental state and environmental flows: The need to reinvent the nation state". In G. Spaargaren, A. P. J. Mol, and F. H. Buttel (Eds.), *Governing Environmental Flows: Global Challenges to Social Theory*, Cambridge, Mass.: MIT, 83–106

Jänicke, M., and H. Weidner (Eds.) (1997), *National Environmental Policies: A Comparative Study of Capacity-Building*, Berlin: Springer

Jänicke, M., M. Binder, et al. (1997), "'Dirty industries': Patterns of change in industrial countries", *Environmental and Resource Economics*, 9, pp. 467–491

Jänicke, M., H. Mönch, M. Binder, et al. (1992), *Umweltentlastung durch industriellen Strukturwandel? Eine explorative Studie über 32 Industrieländer (1970 bis 1990)*, Berlin: Sigma

Jänicke, M. (1993), "Über ökologische und politieke Modernisierungen", *Zeitschrift für Umweltpolitik und Umweltrecht* 2, pp. 159–175

Jasanoff, S. (1996), "The dilemma of environmental democracy", *Issues in Science and Technology*, 13(1), 63–70

Jessop, B. (1990), *State Theory: Putting Capitalist States in Their Place*, University Park: Pennsylvania State University Press

Jimenez-Beltran, D. (1995), "The European environment and cyberspace", Speech to the UNEP Seminar *The Role of the Electronic Highway*, 1 September 1995, Arendal, Norway

Jinyang Zhang (2004), *Analyzing and Evaluating Water Quality Management Policy: A Case Study of Guangzhou City, China*, Wageningen: Wageningen University (M.Sc. thesis)

Jong, W. de (2005), "The power and limits of media-based international oppositional politics – a case study: The Brent Spar conflict". In W. de Jong, M. Shaw, and N. Stammers (Eds.), *Global Activism, Global Media*, London/Ann Arbor: Pluto Press, pp. 110–124

Jordan, A. and A. Schout (2006), *The Coordination of the European Union: Exploring the Capacities of Networked Governance*, Oxford: Oxford University Press

Kamarck, E. C., and J. S. Nye Jr. (Eds.) (1999), *Democracy.com: Governance in a Networked World*, Hollis, N.H.: Hollis Publishing

Karkkainen, B. C. (2001), "Information as environmental regulation: TRI and performance benchmarking, precursors to a new paradigm?", *Georgetown Law Journal*, 89(2), 259–370

Katsuno, M. (2005), *Status and Overview of Official ICT Indicators for China*, STI Working Paper No. 5, Paris: OECD

Kaufman, V. (2002), *Re-thinking Mobility: Contemporary Sociology*, Aldershot, U.K.: Ashgate

Keck, M. E., and K. Sikkink (1998), *Activists beyond Borders: Advocacy Networks in International Politics*, Ithaca, N.Y.: Cornell University Press

Keeley, J. (2006), "Balancing technological innovation and environmental regulation: An analysis of Chinese agricultural biotechnology governance", *Environmental Politics*, 15(2), 293–309

Kersbergen, K. van, and F. van Waarden (2004), "Governance as a bridge between disciplines: Cross-disciplinary inspiration regarding shifts in governance and problems of governability, accountability and legitimacy", *European Journal of Political Research*, 43, 143–171

Kerwer, D. (2005), "Rules that many use: Standards and global regulation", *Governance: An International Journal of Policy, Administration, and Institutions*, 18(4), 611–632

Kesselring, S. (2006), "Pioneering mobilities: New patterns of movement and motility in a mobile world", *Environment and Planning A*, 38(2), 269–279

Klandermans, B. (1986), "New social movements and resource mobilization: The European and the American approach", *International Journal of Mass Emergencies and Disaster*, 4 (2), 13–39

Klein, N. (2000), *No Logo*, London: Flamingo

Kleindorfer, P. R., and E. W. Orts (1999), "Informational regulation of environmental risks", *Risk Analysis*, 18, 155–170

Kloppenburg, S. (2005), *New Inequalities in an Era of Globalisation: Mobilities and In- and Exclusions*, Wageningen: Wageningen University (Master thesis)

Kluver, R. (2005), "The architecture of control: A Chinese strategy for e-governance", *Journal of Public Policy*, 25(1), pp. 75–97

Kluver, R., and Y. Chen (2003), *The Internet in China: Breaching Disciplinary Borders in Academic Research*, paper presented at the International Communications Association, San Diego, Calif.

Kobrin, S. J. (2001), "Territoriality and the governance of cyberspace", *Journal of International Business Studies*, 32(4), pp. 687–704

Kogut, B. (Ed.), *The Global Internet Economy*, Cambridge, Mass.: MIT

Konar, S., and M. A. Cohen (1997), "Information as regulation: The effect of community right to-know law on toxic emissions", *Journal of Environmental Economics and Management*, 32, 109–124

Kooiman, J. (2003), *Governing as Governance*, London: Sage

Kortbech-Olesen, R. (2003), "Market". In M. Yussefi and H. Willer. (Eds.), *The World of Organic Agriculture*, Tholey-Theley, Germany: IFAOM, pp. 21–26

Krut, R., and H. Gleckman (1998), *ISO 14001: A Missed Opportunity for Sustainable Global Industrial Development*, London: Earthscan

Kumar, K. (1995), *From Post-industrial to Post-modern Society: New Theories of the Contemporary World*, Oxford: Blackwell

Lacey, C., and D. Longman (1993), "The press and public access to the environment and development debate", *Sociological Review*, 41(2), 207–243

Lagerkvist, J. (2005), "The techno-cadre's dream: Administrative reform by electronic governance in China today?", *China Information*, 19(2), 189–216

Lash, S. (2002), *Critique of Information*, London: Sage

Lash, S. (2005), "*Lebenssoziologie*: George Simmel in the information age", *Theory, Culture & Society*, 22(3), 1–23

Lash, S., and J. Urry (1987), *The End of Organized Capitalism*, Oxford: Blackwell

Lash, S., and J. Urry (1994), *The Economies of Signs and Spaces*, London: Sage

Latham, K. (2000), "Nothing but the truth: News media, power and hegemony in South China", *China Quarterly*, 163, 633–654

Latour, B. (2004), *Politics of Nature: How to Bring the Sciences into Democracy*, Cambridge, Mass.: Harvard University Press

Le Van Khoa (2006), *Greening Small and Medium-Sized Enterprises: Evaluating Environmental Policy in Viet Nam*, Wageningen: Wageningen University (Ph.D. dissertation)

Leadbeater, C. (1999), *Living on Thin Air: The New Economy*, New York: Viking

Lee, Y.-S. F. (2005), "Public environmental consciousness in China: Early empirical evidence". In K. A. Day. (Ed.), *China's Environment and the Challenge of Sustainable Development*, Armonk, N.Y.: M.E. Sharp, pp. 60–65

Lemos, M. C., and A. Agrawal (2006), "Environmental governance", *Annual Review of Environment and Resources*, 31, 3.1–3.29

Leroy, P., and J. van Tatenhove (2000), "New policy arrangements in environmental politics: The relevance of political and ecological modernization". In G. Spaargaren, A. P. J. Mol, and F. Buttel (Eds.), *Environment and Global Modernity*, London: Sage

LEWIS Public Relations (2007), *The Business Value of Blogging*, Eindhoven: LEWIS Benelux

Leydesdorff, L. (2002), "May there be a 'socionomy' beyond 'sociology'?", *Scipolicy – The Journal of Science and Health Policy*, 2 (http://home.att.net/Scipolicy/index.htm)

Li Junhui (2005), *The Position of Chinese Newspapers in the Framing of Environmental Issues*, Wageningen: Wageningen University (Master of Science thesis)

Lieberthal, K. (1995), *Governing China: From Revolution through Reform*, New York: W.W. Norton

Lipschutz, R., and K. Conca (Eds.) (1993), *The State and Social Power in Global Environmental Politics*, New York: Columbia University Press, pp. 1–13

Litfin, K. (Ed.) (1998), *The Greening of Sovereignty in World Politics*, Cambridge, Mass.: MIT Press, pp. 273–297

Lo, C. W.-H., and S.-Y. Tang (2006), "Institutional reform, economic changes, and local environmental management in China: The case of Guangdong Province", *Environmental Politics*, 15(2), 190–210

Lomborg, B. (1998), *The Skeptical Environmentalist: Measuring the Real State of the World*, Cambridge: Cambridge University Press

Lyon, D. (1988), *The Information Society: Issues and Illusions*, Cambridge, U.K.: Polity

Lyon, D. (2001), *Surveillance Society: Monitoring Everyday Life*, Buckingham, U.K.: Open University Press

Ma, X., and L. Ortolano (2000), *Environmental Regulation in China: Institutions, Enforcement, and Compliance*. Lanham, Md.: Rowman & Littlefield

Machlup, F. (1962), *The Production and Distribution of Knowledge in the United States*, Princeton NJ: Princeton University Press

Macnaghten, P., and M. Jacobs (1997), "Public identification with sustainable development: Investigating cultural barriers to participation", *Global Environmental Change*, 7(1), 5–24

Magat, W. A., and W. K. Viscusi (1992), *Informational Approaches to Regulation*, Cambridge, Mass.: MIT

Majone, G. (1989), *Evidence, Argument, and Persuasion in the Policy Process*, London and New Haven, Conn.: Yale University Press

Manning, P. (2001), *News and News Sources: A Critical Introduction*, London: Sage

Markard, J., et al. (2003), "Disclosure of electricity products – lessons from consumer research as guidance for energy policy", *Energy Policy*, 31, 1459–1474

Martens, S. (2006), "Public participation with Chinese characteristics: Citizen-consumers in China's environmental management", *Environmental Politics*, 15(2), pp. 211–230

Marvin, S., H. Chappells, et al. (1999), "Pathways of Smart Metering Development: Shaping Environmental Innovation", *Computers, Environment and Urban Systems*, 23, pp. 109–126

Massey, D. (1994), *Space, Place and Gender*, Minneapolis: University of Minnesota Press

Mathiesen, T. (1997), "The viewer society: Michel Foucault's 'Panopticon' revisited", *Theoretical Criminology*, 1(2), pp. 215–234

Matten, D. (2003), "Symbolic politics in environmental regulation: Corporate strategic responses", *Business Strategy and the Environment*, 12, pp. 215–226

Matthews, A. S. (2001), *The Environmental Implications of the Growth of the Informaion and Technology Sector*, Paris: OECD (unpublished manuscript, presented 16 April 2001 at the OECD Environment Directorate Workshop)

McBeath, G. and T.-K. Leng (2006), *Governance of biodiversity conservation in China and Taiwan*, Cheltenham: Edward Elgar

McCarthy, D., and M. N. Zald (1980), "Social movement industries: Competition and cooperation among movement organizations", *Research in Social Movements, Conflicts and Change*, 3, pp. 1–20

McCarthy, J. D., and M. N. Zald (Eds.) (1977a), *The Dynamics of Social Movements*, New York: Winthrop Publishers

McCarthy, J. D., and M. N. Zald (1977b), "Resource mobilization and social movements: A partial theory", *American Journal of Sociology*, 82 (May), 1212–1239

McCright, A. M., and R. E. Dunlap (2000), "Challenging global warming as a social problem: An analysis of the conservative movement's counter-claims", *Social Problems*, 47(4), 499–522

McCright, A. M., and R. E. Dunlap (2003), "Defeating Kyoto: The conservative movement's impact on US climate change policy", *Social Problems*, 50(3), 348–373

McKibben, B. (1993), *The Age of Missing Information*, New York: Penguin Plume

McLuhman, M., and Q. Fiore (1967), *The Medium Is the Message*, Harmondsworth: Penguin

McMichael, P. (1996), "Globalisation: Myths and realities", *Rural Sociology*, 61(1), 25–55

McNaghten, P., and J. Urry (1998), *Contested Natures*, London: Sage

McNeil, G., and D. Hathaway (2005), "Green labeling and energy efficiency in China", *China Environment Series*, 7, 72–73

McPhail, T. L. (2006), *Global Communication: Theories, Stakeholders and Trends*, Malden, Mass.: Blackwell

Metz, B., A. P. J. Mol, M. Andersson, M. Berk, J. van Minnen and W. Tuinstra (2003), "Possible strategies for far reaching emission reductions: The COOL project". In M. Kok, J. Gupta and E. van Ierland. (Eds.), *Issues in International Climate Policy*, Cheltenham: Edward Elgar, pp. 263–284

Meulen, B. van der, and M. van der Velde (2004), *Food Safety Law in the European Union: An Introduction*, Wageningen: Wageningen Academic Publishers

Meyrowitz, J. (1985), *No Sense of Place: The Impact of Electronic Media on Social Behavior*, New York: Oxford University Press

Miles, I. (1988), *Home Informatics: Information Technology and the Transformation of Everyday Life*, London: Pinter Publications

Miller, D., and D. Slater (2000), *The Internet: An Ethnographic Approach*, London: Berg

Milton, K. (Ed.) (1994), *Environmentalism. The View from Anthropology*, London/New York: Routledge

Minoli, D. M., and J. N. B. Bell (2003), "Insurance as an alternative environmental regulator: Findings from a retrospective pollution claims survey", *Business Strategy and the Environment*, 12, 107–117

Mol, A. P. J. (1995), *The Refinement of Production: Ecological Modernization Theory and the Chemical Industry*, Utrecht: International Books

Mol, A. P. J. (1996), "Ecological modernisation and institutional reflexivity. Environmental reform in the late modern age", *Environmental Politics*, 5(2), 302–323

Mol, A. P. J. (2000) "The environmental movement in an Age of ecological modernisation", *Geoforum*, 31(1), 45–56

Mol, A. P. J. (2001), *Globalization and Environmental Reform: The Ecological Modernization of the Global Economy*. Cambridge, Mass.: MIT Press

Mol, A. P. J. (2003), "Joint environmental policy-making in Europe: Between deregulation and political modernization", *Society and Natural Resources*, 16(4), 335–348

Mol, A. P. J. (2006a), "Environmental governance in the information age: The emergence of informational governance", *Environment and Planning C*, 24(4), 497–514

Mol, A. P. J. (2006b), "From environmental sociologies to environmental sociology? A comparison of U.S. and European environmental sociology", *Organization & Environment*, 19(1), 5–27

Mol, A. P. J., and F. H. Buttel (Eds.) (2002), *The Environmental State under Pressure*, Amsterdam/New York: Elsevier/JAI

Mol, A. P. J., and N. Carter (2006), "China's environmental governance in transition", *Environmental Politics*, 15(2), 149–170

Mol, A. P. J., V. Lauber and J. D. Liefferink (Eds.) (2000), *The Voluntary Approach to Environmental Policy: Joint Environmental Policy-Making in the EU and Selected Member States*, Oxford: Oxford University Press

Mol, A. P. J., and Y. Liu (2005), "Institutionalising cleaner production in China: The Cleaner Production Promotion Law", *International Journal of Environment and Sustainable Development*, 4(3), 227–245

Mol, A. P. J., J. H. Mol and B.van Vliet (Eds.) (2004), *Suriname Schoon genoeg? Hulpbronnengebruik en Milieubeheer in een klein Amazoneland*, Utrecht: International Books

Mol, A. P. J., and D. A. Sonnenfeld (Eds.) (2000), *Ecological Modernization around the World: Perspectives and Critical Debates*, London: Frank Cass

Mol, A. P. J., and G. Spaargaren (2002), "Ecological modernization and the environmental state". In A. P. J. Mol and F. H. Buttel (Eds.), *The Environmental State under Pressure*, Amsterdam/Oxford: Elsevier, pp. 33–52

Mol, A. P. J., and G. Spaargaren (2004), "Ecological modernization and consumption: A reply", *Society and Natural Resources*, 17, 261–265

Mol, A. P. J., and G. Spaargaren (2000), "Ecological Modernization Theory in debate: a review", *Environmental Politics* 9(1), 17–49

Mol, A. P. J., and G. Spaargaren (2005), "From Additions and Withdrawals to Environmental Flows. Reframing Debates in the Environmental Social Sciences" *Organization & Environment* (18)1, 91–107

Mol, A. P. J., and G. Spaargaren (2006), "Towards a sociology of environmental flows: A new agenda for twenty-first-century environmenal sociology". In G. Spaargaren, A. P. J. Mol, and F. H. Buttel. (Eds.), *Governing Environmental Flows: Global Challenges for Social Theory*, Cambridge, Mass.: MIT, pp. 39–83

Mol, A. P. J. and J. C. L. van Buuren (Eds.) (2003), *Greening Industrialization in Asian Transitional Economies: China and Vietnam*, Lanham: Rowman & Littlefield

Morris, D. (2004), "Globalization and Media Democracy: The Case of Indymedia". In D. Schuler and P. Day (Eds.), *Shaping the Network Society: The New Role of Civil Society in Cyberspace*, Cambridge, Mass.: MIT, pp. 325–352

Morris-Suzuki, T. (1984), "Robots and capitalism", *New Left Review*, 147, 109–121

Moxen, J. and A. McCulloch (1999), "Organizing the dissemination of environmental information: lessons from Scotland", *Journal of Environmental Policy and Planning*, 1, 2, pp. 155–165

MPI (Ministry of Planning and Investment) (2001), *Public Disclosure of Environmental Performance in Development Planning: Lessons from HCM's Black and Green Book*, Hanoi: Ministry of Planning and Investment

Murdoch, J., and M. Miele (1999), "'Back to nature': Changing 'world of production' in the food sector", *Sociologia Ruralis*, 39(4), 465–483

Murphy, D. F. and J. Bendell (1997), *In the Company of Partners. Business, Environmental Groups and Sustainable Development Post-Rio*. Bristol: The Policy Press

Naisbitt, J. (1984), *Megatrend: Ten New Directions Transforming Our Lives*, New York: Warner Books

Natan, T. E., and C. G. Miller (1998), "Are toxics release inventory reductions real?", *Environmental Science and Technology*, 32(15), 369–374

NEWGOV (2004), *Integrated Project "New Modes of Governance" – Description of Work*, Florence: European University Institute

Noble, D. F. (1986), *Forces of Production: A Social History of Industrial Automation*, New York: Oxford University Press

Noe, P., F. R. Anderson, S. A. Shapiro, et al. (2003), "Learning to live with the Data Quality Act", *Environmental Law Reporter*, 33, 10224–10236

Nogueira, A. (2002), "The birth and promise of the Indymedia revolution". In B. Shepard and R. Hayduk (Eds.), *From ACT UP to the WT: Urban Protest and Community Building in the Era of Globalization*, London: Verso, pp. 290–297

Norberg-Bohm, V., and T. de Bruijn (2005), "Conclusions: Lessons for the design and use of voluntary, collaborative, and information-based approaches to environmental policy". In T. de Bruijn and V. Norberg-Bohm (Eds.), *Industrial Transformation: Environmental Policy Innovation in the United States and Europe*, Cambridge, Mass.: MIT, pp. 361–387

Notes from Nowhere (Ed.) (2003), *We Are Everywhere: The Irresistible Rise of Global Anti-capitalism*, London: Verso

Nouwt, S., C. Prins, B. van der Meulen, and M. Lugt (2004), *Position Paper: Transparency in Network Chains – Legal Dimensions*, Tilburg/Wageningen: Tilburg University/Wageningen University (http://www.klict.org/frames.asp)

Noveck, B. S. (2003), "Designing deliberative democracy in cyberspace: The role of the cyber-lawyer", *Boston University Journal of Science & Technology Law*, 9(1), 1–71

Nthunya, E. (2002), "The role of information in environmental management and governance in Lesotho", *Local Environment*, 7(2), 135–148

Nyquist, S. (2003), "The legislation of environmental disclosures in three Nordic countries – a comparison", *Business Strategy and the Environment*, 12, 12–25

O'Neill, J. (2001) "Representing people, representing nature, representing the world", *Environment and Planning C: Government and Policy*, 19, 483–500

O'Donovan, G. (1999), "Managing legitimacy through increased corporate environmental reporting: An exploratory study", *Interdisciplinary Environmental Review*, 1(1), 63–99

O'Donovan, G. (2002), "Environmental disclosures in the annual report: Extending the applicability and predictive power of legitimacy theory", *Accounting, Auditing and Accountability Journal*, 15(3), 344–371

O'Rourke, D. (2004), *Community-Driven Regulation: Balancing Development and the Environment in Vietnam*, Cambridge, Mass.: MIT

Offe, C. (1985), "New social movements: Challenging the boundaries of institutional politics", *Social Research*, 52(4), 817–868

Ohmae, K. (1995), *The End of the Nation State: The Rise of Regional Economies*, New York: Simon & Schuster

Oosterveer, P. (2005), *Global Food Governance*, Wageningen: Wageningen University (Ph.D. dissertation)

Orts, E. W. (1995), "A reflexive model of environmental regulation", *Business Ethics Quarterly*, 5(4), 779–794

Overdevest, C. (2005), "Treadmill politics, information politics, and public policy: Toward a political economy of information", *Organization & Environment*, 18(1), 72–90

Owens, S. (2000), "Engaging the public: Information and deliberation in environmental policy", *Environment and Planning A*, 32, 1141–1148

Paehlke, R. C. (2003), *Democracy's Dilemma: Environment, Social Equity and the Global Economy*, Cambridge, Mass.: MIT

Park, J., and N. Roome (Eds.) (2002), *Ecology of the New Economy: Sustainable Transformation of Global Information Communication and Electronics Industries*, Sheffield, U.K.: Greenleaf Publishing

Paterson, M., and S. Dalby (2006), "Empire's ecological tyreprints", *Environmental Politics*, 15(1), 1–22

Pattberg, P. (2005), "The institutionalization of private governance: How business and nonprofit organizations agree on transnational rules", *Governance: An International Journal of Policy, Administration, and Institutions*, 18(4), 589–610

Patten, D. M. (1992), "Intra-industry environmental disclosures in response to the Alaskan oil spill: A note on legitimacy theory", *Accounting, Organizations and Society*, 17(5), 471–575

Pellizzoni, L. (2003), "Uncertainty and participatory democracy", *Environmental Values*, 12, 195–224

Pellow, D. N., A. S. Weinberg and A. Schnaiberg (2000), "Putting ecological modernization to the test: Accounting for recycling's promises and performance", *Environmental Politics*, 9(1), 109–137

Pepper, D. (1984), *The Roots of Modern Environmentalism*, London: Croom Helm

Pepper, D. (1999), "Ecological modernisation or the 'ideal model' of sustainable development? Questions prompted at Europe's periphery", *Environmental Politics*, 8(4), 1–34

Perkin, H. (1989), *The Rise of Professional Society: Britain since 1880*, London: Routledge

Peters, M. (2004), "Why supply chain transparency is a consumer concern". In G. J. Hofstede, H. Schepers, L. Spaans-Dijkstra, et al. (Eds.), *Hide or Confide? The Dilemma of Transparency*, Den Hague: Reed Business Information, pp. 173–182

Petkova, E., C. Maurer, N. Henninger and F. Irwin (2002), *Closing the Gap: Information, Participation, and Justice in Decision-Making for the Environment*, Washington, D.C.: World Resources Institute

Pham Minh Hai (2005), *Environmental Monitoring System in Hanoi*, Wageningen: Wageningen University (M.Sc. thesis)

Phung Thuy Phuong, and A. P. J. Mol (2004), "Communities as informal regulators: New arrangements in industrial pollution control in Viet Nam", *Journal of Risk Research*, 7(4), pp. 431–444

Pickerill, J. (2000), "Spreading the green word? Using the Internet for environmental campaigning", *ECOS*, 21(1), 14–24

Pickerill, J. (2001), "Environmentalists' Internet activism in Britain", *Peace Review*, 13(3), pp. 365–370

Pickerill, J. (2003a), *Cyberprotest: Environmental Activism On Line*, Manchester: Manchester University Press

Pickerill, J. (2003b), *Out in the Open: Indymedia Networks in Australia*, paper presented at the symposium *Information, Communication and Society*, Oxford, Oxford University, September 2003

Pickerill, J. (2004), "Rethinking political participation: Experiments in Internet activism in Australia and Britain". In R. Gibson, A. Roemmele, and S. Wards (Eds.), *Electronic Democracy: Mobilization, Organization and Participation via New ICTs*, London: Routledge, pp. 170–193

Pierre, J., and B. G. Peters. 2005. *Governing complex societies: Trajectories and scenarios*. Basingstoke, UK: Palgrave.

Ping Song, S. (2005), "Atypical environmental NGOs in Guangdong, China", *China Environment Series*, 7, 65–71

Pirages, D., and K. Cousins (Eds.) (2005), *From Resource Scarcity to Ecological Security: Exploring New Limits to Growth*, Cambridge, Mass.: MIT

Plepys, A. (2002), "The grey side of ICT", *Environmental Impact Assessment Review*, 22, 509–523

Pollitt, C. (2001), "Clarifying convergence: Striking similarities and durable differences in public management reform", *Public Management Review*, 3, 471–492

Ponte, S. (2006), *Ecolabels and Fish Trade: Marine Stewardship Council Certification and the South African Hake Industry*, Tarlac Working Paper 9/2006, Copenhagen: Danish Institute for International Studies

Ponte, S., and P. Gibbon (2005), "Quality standards, conventions and the governance of global value chains", *Economy and Society*, 34(1), 1–31

Porat, M. (1977), *The Information Economy*. Washington, DC: US Department of Commerce

Porter, M. E., and C. M. H. Ketels (2003), *UK Competitiveness: Moving to the Next Stage*, DTI Economics Papers No. 3

Powel, M. R. (1999), *Science at EPA: Information in the Regulatory Process*, Washington, D.C.: Resources for the Future

Powell, M. R. (1999). *Science at EPA: information in the regulatory process*. Washington DC, Resources for the future

Power, M. (1997), *The Audit Society: Rituals of Verification*, Oxford: Oxford University Press

Presas, L. M. S. (2005), *Transnational Buildings in Local Environments*, Burlington, Vt.: Ashgate

Presas, L. M. S., and A. P. J. Mol (2006), "Greening transnational buildings: In-between global flows and local places". In G. Spaargaren, A. P. J. Mol and F. H. Buttel (Eds.), *Governing Environmental Flows: Global Challenges for Social Theory*, Cambridge, Mass.: MIT, 303–326

Prévost, Y., and P. Gilruth (1999), *Environmental Information Systems in Sub-Saharan Africa*, EIS-SSA project (http://www.grida.no/eis-ssa/index.htm)

Prittwitz, V. von (Ed.) (1993), *Umweltpolitik als Modernisirungsprozeß: Politikwissenschaftliche Umweltforschung und -lehre in der Bundesrepublik*, Opladen: Leske + Budrich

Putnam, R. (2000), *Bowling Alone: The Collapse and Revival of American Community*, New York: Simon & Schuster

Qian, Y., et al. (2001), *Development Strategy for Organic Food Industry in China* (http://www.bioone.org/pdfserv/i0044-744-030-07-0450.pdf)

Qing, D., and E. Vermeer (1999), "Do good work, but do not offend the 'old communists': Recent activities of China's non-governmental environmental protection organizations and individuals." In W. Draguhn and R. Ash (Eds.), *China's Economic Security*, Richmond: Curzon Press, pp. 142–162

Qiu, J. L., and N. Hachigian (2004), *A New Long March: E-Government in China*, Paris: OECD

Raikes, P., M. Friis Jensen, and S. Ponte (2000), "Global commodity chain analysis and the French filière approach: Comparison and critique", *Economy and Society*, 29(3), 390–417

Raynolds, L. T. (2004), "The globalization of organic agro-food networks", *World Development*, 32(5), 725–743

Richards, D. J., B. R. Allenby and W. D. Compton (Eds.) (2001), *Information Systems and the Environment*, Washington, D.C.: National Academy Press

Rifkin, J. (2000), *The Age of Access. How the Shift from Ownership to Access is Transforming Modern Life*, London: Penguin

Robinson, P. (2002), *The CNN Effect: The Myth of News Media, Foreign Policy and Intervention*, London: Routledge

Rock, M. T. (2002), "Getting into the environment game: Integrating environmental and economic policy-making in China and Taiwan", *American Behavioral Scientist*, 45(9), pp. 1435–1455

Roe, D. (2002), "Toxic chemical control policy: Three unabsorbed facts" *Environmental Law Reporter*, 32, 10232–10239

Rogers, R. (2002), "Operating issue networks on the web", *Science as Culture*, 11(2), 191–213

Romm, J. (1999), *The Internet Economy and Global Warming* (available at www.coolcompanies.org/paper/.cfn)

Rosenau, J. N., and E.-O. Czempiel (Eds.) (1992), *Governance without Government: Order and Change in World Politics*, Cambridge: Cambridge University Press

Rosenbaum, W. A. (2000), "Escaping the 'battered agency syndrome': EPA's gamble with regulatory reinvention". In N. J. Vig and M. E. Kraft (Eds.), *Environmental Policy*, 4th ed., Washington, D.C.: CQ Press, pp. 165–189

Roszak, T. (1986), *The Cult of Information: The Folklore of Computers and the True Art of Thinking*, Cambridge: Lutterworth

Ruiter, de W. (1988), "Het postmoderne tijdperk", *Wetenschap en Samenleving*, 40(4), 3–11

Saar, S. and V. Thomas (2002), "Towards trash that thinks: Product tags for environmental management", *Journal of Industrial Ecology*, 6(2), 133–146

Sagoff, M. (1988), *The Economy of the Earth*, Cambridge: Cambridge University Press

Salzman, J. (1999), "Beyond the smokestack: Environmental protection in the service economy", *UCLA Law Review*, 47, 411–489

Sand, P. H. (2002), "*The right to know: Environmental information disclosure by government and industry,*" paper presented at the conference *Human Dimensions of Global Environmental Change: Knowledge for the Sustainability Transition*, 7 December 2002, Berlin, Germany

Sandywell, B. (2003), "Metacritique of information: On Scott Lash's *Critique of Information*", *Theory, Culture & Society*, 20(1), 109–122

Sassen, S. (1994), *Cities in a World Economy*, Thousand Oaks, Calif: Pine Forge Press

Saxby, S. (1990), *The Age of Information: The Past Development and Future Significance of Computing and Communications*, Basingstoke and London: Macmillan

Schienke, E. (2001a), *Bill Pease; An Original Developer of Scorecard.org*, CECS Working Interviews, Troy, N.Y.: Center for Ethics in Complex Systems RPI

Schienke, E. (2001b), *Michael Stanley-Jones: Scorecard and Toxics Information*, CECS Working Interviews, Troy, N.Y.: Center for Ethics in Complex Systems RPI

Schiller, H. I. (1969), *Mass Communications and American Empire*, New York: Augustus M. Kelly

Schiller, H. I. (1981), *Who Knows: Information in the Age of the Fortune 500*, Norwood, N.J.: Ablex

Schiller, H. I. (1985), "Beneficiaries and Victims of the Information Age: The Systematic Diminution of the Public Supply of Meaningful Information", *Papers in Comparative Studies* 4, pp. 185–192

Schiller, H. I. (1989), *Culture Inc.: The Corporate Takeover of Public Expression*, New York: Oxford University Press

Schlosberg, D. and J. Dryzek (2002). "Digital Democracy: Authentic or Virtual." *Organization and Environment* 15, No. 3: 327–330.

Schnaiberg, A. (1980), *The Environment: From Surplus to Scarcity*, Oxford and New York: Oxford University Press

Schnaiberg, A., A. S. Weinberg and D. N. Pellow (2002), "The treadmill of production and the environmental state". In A. P. J. Mol and F. H. Buttel (Eds.), *The Environmental State under Pressure*, London: JAI/Elsevier, pp. 15–32

Scholte, J. A. (2000), *Globalization: A Critical Introduction*, New York: Palgrave

Schuler, D., and P. Day (Eds.) (2004), *Shaping the Network Society. The New Role of Civil Society in Cyberspace*, Cambridge, Mass.: MIT

Schumacher, E. F. (1973), *Small Is Beautiful*, London: Blond & Briggs

Scoble, R., and S. Israel (2006), *Naked Conversations: How Blogs Changed the Way in which Businesses Talk with Customers*, Hoboken, N.J.: Wiley

Scott, A., and J. Street (2001), "From media politics to e-protest? The use of popular culture and new media in parties and social movements". In F. Webster. (Ed.), *Culture and Politics in the Information Age: A New Politics?*, London: Routledge, pp. 32–51

Seidman, G. (2005), "'Stateless' regulation and consumer pressure: Historical experiences of transnational corporate monitoring". In F. H. Buttel and P. McMichael (Eds.), *New Directions in the Sociology of Global Development*, Amsterdam: Elsevier, pp. 175–207

Shang Hongbo (2004), *The Tactics and Strategies Used by Collective Pollution Victims to Defend Their Rights and Interests in Industrial Pollution Conflicts in China*, Wageningen: Wageningen University (M.Sc. thesis)

Shi, H., and L. Zhang (2006), "Governing China's rapid industrialization for the environment: Challenges and innovations", *Environmental Politics*, 15(2), 272–293

Shove, E. (1997), "Revealing the invisible: Sociology, energy and the environment". In M. Redclift and G. Woodgate (Eds.), *The International Handbook of Environmental Sociology*, Cheltenham: Edward Elgar, pp. 261–273

Shulman, S. W. (2004), *The Internet Still Might (but Probably Won't) Change Everything: Stakeholder Views on the Future of Electronic Rulemaking*, Pittsburgh: University of Pittsburgh

Simmons, S. E. (2002), "E-work and sustainability part II: The basis of Immaterialisation in lifestyle change; Assessing the impact of lifestyle change(LCA-I); What happens after immaterialisation" *The European Journal of Teleworking* 8(2). (http://www.immaterialisation.org)

Simonsen, K. (2004), "Networks, flows and fluids – reimagining spatial analysis?", *Environment and Planning A*, 36, 1333–1337

Sinclair, G. (2002), "The Internet in China: Information revolution or authoritarian solution?" (http://www.geocities.com/gelaige79/intchin.pdf)

Singer, S. F. (2000), "Where politics trumps science", *Regulation*, 23(1), 69–72

Sinton, J. E., and D. G. Fridley (2001), "Hot air and cold water: The unexpected fall in China's energy use", *China Environment Series*, 4, 3–20

Sinton, J. E., and D. G. Fridley (2003), "Comments on recent energy statistics from China", *Sinosphere Journal*, 6(2), pp. 6–11

Slob, A., and M. van Lieshout (2002), "The contribution of information and communication technologies to the transition towards a climate-neutral society". In M. Kok, W. Vermeulen, A. Faaij, and D. de Jager (Eds.), *Global Warming & Social Innovation: The Challenge of a Climate-Neutral Society*, London: Earthscan, pp. 143–159

Slosberg, D., and J. S. Dryzek (2002), "Digital democracy: Authentic or virtual?", *Organization & Environment*, 15(3), 332–335

Smith, T., D. A. Sonnenfeld and D. N. Pellow (Eds.) (2006), *Challenging the Chip: Labor Rights and Environmental Justice in the Global Electronics Industry*, Philadelphia: Temple University Press

Sonnenfeld, D. A., and A. P. J. Mol (2002), "Globalization and the Transformation of Environmental Governance: An Introduction", *American Behavioral Scientist*, 45(9), 1318–1339

Sonnenfeld, D. A., and A. P. J. Mol (2006), "Environmental reform in Asia: Comparisons, challenges, next steps", *Journal of Environment and Development*, 15(2), 112–137

Spaargaren, G. (1987), "Environment and society: Environmental sociology in the Netherlands", *The Netherlands' Journal of Sociology*, 23(1), 54–72

Spaargaren, G. (1997), *The Ecological Modernisation of Production and Consumption: Essays in Environmental Sociology*, Wageningen: Wageningen Agricultural University (Ph.D. dissertation)

Spaargaren, G. (2003), "Sustainable consumption: A theoretical and environmental policy perspective", *Society and Natural Resources*, 16(8), 687–702

Sparks, C. (2005), "Media and the global public sphere: An evaluative approach". In W. de Jong, M. Shaw, and N. Stammers (Eds.), *Global Activism, Global Media*, London/Ann Arbor: Pluto Press, pp. 34–49

Staats, H. J., and P. Harland (1995), *The Eco-team Program in the Netherlands, Study 4: A Longitudinal Study of the Effects on Environmental Behavior and Its Psychological Background*, Leiden: University of Leiden, Centre for Energy and Environmental Research

Stalley, P., and D. Yang (2006), "An emerging environmental movement in China?", *China Quarterly*, 186, 333–356

Stehr, N. (1994), *Knowledge Societies: The Transformation of Labor, Property and Knowledge in Contemporary Societies*, London: Sage

Stephan, M. (2002), "Environmental information disclosure programs: They work, but why?", *Social Science Quarterly*, 83(1), 190–205

Stevens, H. (1999), *The Institutional Position of Sea Ports: An International Comparison*, Dordrecht: Kluwer/Springer

Stevenson, N. (2001), "The future of public media cultures: Morality, ethics and ambivalence". In F. Webster (Ed.), *Culture and Politics in the Information Age: A New Politics?*, London: Routledge, pp. 63–80

Stevis, D., and H. Bruyninckx (2006), "Looking through the state at environmental flows and governance". In G. Spaargaren, A. P. J. Mol, and F. H. Buttel (Eds.), *Governing Environmental Flows: Global Challenges for Social Theory*, Cambridge, Mass.: MIT, pp. 107–137

Stewart, R. B. (2001), "A new generation of environmental regulation?", *Capital University Law Review*, 29, 21–182

Stockholm Environment Institute (2002), *Making Green Development a Choice: China Human Development Report*, Oxford: Oxford University Press

Stonier, T. (1983), " The Microelectronic Revolution, Soviet Political Structure, and the Future of East/West Relations", *The Political Quarterly* 54 (2), 137–151

Stretesky, P. B., and J. Gabriel (2005), "Self-policing and the environment: Predicting self-disclosure of Clean Air Act violations under the U.S.

Environmental Protection Agency's audit policy", *Society and Natural Resources*, 18(10), 871–887

Sunstein, C. R. (1999), "Informational regulation and informational standing: Akins and beyond", *147 University of Pennsylvania Law Review*, 613, 618–624

Taipale, K. A. (2003), *Information Technology as Agent of Change in Environmental Policy*, Working Paper No. 05–2003, Centre for Advanced Studies (http://www.advancedstudies.org/papers/agentofchange.pdf)

Tarrow, S. (1998), *Power in Movement*, 2nd ed., Cambridge: Cambridge University Press

Tatenhove, J. van, B. Arts and P. Leroy (Eds.) (2000), *Political Modernisation and the Environment: The Renewal of Policy Arrangements*, Dordrecht: Kluwer

Thiers, P. (1999), *Green Food: The Political Economy of Organic Agriculture in China*, (unpublished Ph.D. dissertation, Department of Political Science, University of Oregon, Eugene)

Thompson, G. F. (2004), "Getting to know the knowledge economy: ICTs, networks and governance", *Economy and Society*, 33(4), 562–581

Thompson, J. B. (1995), *The Media and Modernity*, Cambridge, U.K.: Polity

Tickner, J. A., C. Raffensperger and N. Myers (1999), *The Precautionary Principle in Action*, Windsor, N.D.: Science and Environmental Health Network

Tietenberg, T. (1998), "Disclosure strategies for pollution control", *Environmental and Resource Economics*, 11, 587–602

Tietenberg, T., and D. Wheeler (1998), *Empowering the Community: Information Strategies for Pollution Control*, paper delivered at the Environmental Economics Conference, Arlie House, Virginia, October 1998

Toffler, A. (1970), *Future Shock*, New York: Random House

Toffler, A. (1981), *The Third Wave*, New York: Bantam Books

Tomlinson, J. (1999), *Globalization and Culture*, Cambridge, U.K.: Polity

Touraine, A. (1971), *The Post-industrial Society: Tomorrow's Social History; Classes, Conflicts and Culture in the Programmed Society*, New York: Wildwood House

Tracy, M. (2005), "Food, environment and health post-SARS: Corporate expectations and participation", *China Environment Series*, 7, 61–65

Tran Thi My Dieu, Phung Thuy Phuong, J. C. L. van Buuren and Nguyen Trung Viet (2003), "Environmental management for industrial zones in Vietnam". In A. P. J. Mol and J. C. L. van Buuren (Eds.), *Greening Industrialization in Asian Transitional Economies: China and Vietnam*, Lanham, Md.: Lexington, pp. 39–58

Treib, O., H. Bähr, and G. Falkner (2007), "Modes of governance: Towards a conceptual clarification", *Journal of European Public Policy*, 14(1), 1–20

Tulbure, I. (2002), "The information society and the environment: A case study concerning two Internet applications". In B. Stanford-Smith, E. Chiozzy, and M. Edin (Eds.), *Challenges and Achievements in E-Business and E-Work*, conference proceedings of the 2002 eBusiness and eWork Conference, October 16–18 2002, Prague, pp. 125–132 (http://www. tukkk.fi/tutu/terra2000/)

Tumber, H. (2001), "Democracy in the information age: The role of the fourth estate in cyberspace". In F. Webster (Ed.), *Culture and Politics in the Information Age: A New Politics?*, London: Routledge, pp. 17–31

Udall, L. (1998), "The World Bank and public accountability: Has anything changed?". In J. A. Fox and L. D. Brown (Eds.), *The Struggle for Accountability: The World Bank, NGOs, and Grassroots Movements*, Cambridge, Mass.: MIT, pp. 391–436

UNCTAD (2001), *E-Commerce and Development Report 2001*, New York: UN

UNCTAD/WTO (2007), *Internaional Trade Centre – Country Profile Organic Products* (http://www.intracen.org/organics/Country-Profiles-overview.htm, accessed November 2007)

UNESCAP (2002), "National study: China". In *Organic Agriculture and Rural Poverty Alleviation, Potential and Best Practices in Asia*, Economic and Social Commission for Asia and the Pacific of the United Nations (http://www.unescap.org/rural/doc/OA/China.PdF)

UNESCO (2005), *Towards Knowledge Societies*, Paris: UNESCO

Urry, J. (2000), *Sociology beyond Society*, London: Routledge

Urry, J. (2003), *Global Complexity*, Cambridge, U.K.: Polity

Urry, J. (2004), "Introduction: Thinking society anew". In U. Beck and J. Willms (Eds.), *Conversations with Ulrich Beck*, Cambridge, U.K.: Polity, pp. 1–10

Vliet, B., J. M. van (2002), *Greening the Grid: The Ecological Modernisation of Network-Bound Systems*, Wageningen: Wageningen University (Ph.D. dissertation)

Vliet, B. J. M. van (2000), "Monitoring and power". In H. Chappells, M. Klintman, A. L. Linden, et al. (Eds.), *Domestic Consumption, Utility Services and the Environment*, Final DOMUS Report, Wageningen: Wageningen University

Völlink, T. and R. M. Meertens (1999), "De effectiviteit van elektronische feedback over het energie-en waterverbruik door middel van teletekst bij huishoudens". In: R. M. Meertens, R. Vermunt, J. de Wit and J.-F. Ybema, (1999), *Sociale psychologie en haar toepassingen* , Delft Eburon, pp. 97–92

Vorst, J. G. A. J. van der, J. van Beurden and H. Folkerts (2003), *Tracking and Tracing of Food Products – An International Benchmark Study in Food Supply Chains*, Houten, the Netherlands: Rijnconsult

Waart, S. de, and D. Spruyt (2001), *Keur of Willekeur: Een inventarisatie van ideële keurmerken en groene beeldmerken*, Amsterdam: Alternatieve Konsumenten Bond

Walker, R. A. (1985), "Is there a service economy? The changing capitalist division of labor", *Science and Society*, 49, 42–83

Wallerstein, I. (1991), *Geopolitics and Geoculture*, Cambridge: Cambridge University Press

Walton, G. (2001), *China's Golden Shield: Corporations and the Development of Surveillance Technology in the People's Republic of China*, International Centre for Rights and Democracy (http://www.ichrd.ca/)

Wang, H., J. Bi, D. Wheeler, et al. (2002), *Environmental Performance Rating and Disclosure: China's Green-Watch Program*, Policy Research Working Paper No. 2889, Washington, D.C.: World Bank

Wang, Q.-J. (2005), "Transparency in the grey box of China's environmental governance: A case study of print media coverage of an environmental controversy from the Pearl River Delta region", *Journal of Environment & Development*, 14(2), 278–312

Wang Sung, and Xie Yan (Eds.) (Biodiversity Working Group of China Council for International Cooperation on Environment and Development) (2004), *China Species Red List, Vol. I*, Beijing: Higher Education Press

Wanxin Li (2006), "Opening up the floor: Environmental performance information disclosure pilot programs in Zhenjiang and Hohhot", *China Environment Series*, 8, 125–129

Wapner, P. (1996), *Environmental Activism and World Civic Politics*, Albany: State University of New York Press

Wapner, P. (1998), "Reorienting state sovereignty: Rights and responsibilities in the environmental age". In K. Litfin, *The Greening of Sovereignty in World Politics*, Cambridge, Mass.: MIT Press, pp. 273–297

Warkentin, C. (2001), *Reshaping World Politics: NGOs, the Internet and Global Civil Society*, Lanham, Md.: Rowman & Littlefield

Washbourne, N. (2001), "New forms of organizing? Translocalism, networks and campaigning in Friends of the Earth". In F. Webster. (Ed.) (2001), *Culture and Politics in the Information Age: A New Politics?*, London: Routledge, pp. 129–141

Waterman, P. (2005), "Between a political-institutional past and a communicational-networked future? Reflections on the Third World Social Forum, 2003". In W. de Jong, M. Shaw, and N. Stammers (Eds.), *Global Activism, Global Media*, London/Ann Arbor: Pluto Press, pp. 68–83

Water Resources Research Institute (2004), *The Data Quality Act: A Revolution in the Role of Science in Policy Making or a Can of Worms?*, Raleigh: The Water Resources Research Institute (http://www.thecre.com/misc/20040606_worms.htm)

Webster, F. (2002), *Theories of the Information Society*, 2nd ed., London/New York: Routledge

Webster, F. (Ed.) (2001a), *Culture and Politics in the Information Age: A New Politics?*, London: Routledge

Webster, F. (2001b), "A new politics?". In F. Webster (Ed.), *Culture and Politics in the Information Age: A New Politics?*, London: Routledge, pp. 1–13

Webster, F. (Ed.) (2004), *The Information Society Reader*, London/New York: Routledge

Wehling, P. (1992), *Die Moderne als Sozialmythos: Zur Kritik sozialwissenschaftlicher modernisierungstheorien*, Frankfurt/New York: Campus

West, D. (2005), *Global E-Government, 2005*, Providence, R.I.: Brown University (http://www.insidepolitics.org/egovt05int.pdf)

Whitley, R. (1996), "Business systems and global commodity chains: Competing or complementary forms of economic organization?" *Competition and Change* 1, pp. 411–425

Whitley, R. (1999), *Divergent Capitalism: The Social Structuring and Change of Business Systems*, New York and Oxford: Oxford University Press

Whittemore, H. (1990), *CNN: The Inside Story*, Toronto: Little, Brown

Wilenius, M. (2002), "Sustainable development and human capital in the network society: The challenge Europe is facing in the future", *Communication & Cognition*, 35(1–2), 1–24

Wilhelm, A. G. (2000), *Democracy in the Digital Age: Challenges to Political Life in Cyberspace*, New York: Routledge

Wilhelm, A. G. (2004), *Digital Nation: Towards an Inclusive Information Society*, Cambridge, Mass.: MIT

Wilson, E. J., III (2004), *The Information Revolution and Developing Countries*, Cambridge, Mass.: MIT

Wong, W., and E. Welch (2004), "Does e-government promote accountability? A comparative analysis of website openness and government accountability", *Governance: An International Journal of Policy, Administration, and Institutions*, 17(2), 25–297

Wood, D., and R. Schneider (2006), "ToxicDude.com: The Dell campaign". In T. Smith, D. A. Sonnenfeld, and D. N. Pellow (Eds.), *Challenging the Chip: Labor Rights and Environmental Justice in the Global Electronics Industry*, Philadelphia: Temple University Press, pp. 285–297

World Bank (1999), *Greening Industry: New Roles for Communities, Markets and Governments*, Washington, D.C.: World Bank

World Bank (2002), *World Development Report 2002: Building Institutions for Markets*, Washington, D.C.: World Bank

Wright, S. (2004), "Informing, communicating and ICTs in contemporary anti-capitalist movements". In W. van de Donk, B. D. Loader, P. G. Nixon, and D. Rucht (Eds.), *Cyberprotest: New Media, Citizens and Social Movements*, London: Routledge, pp. 77–93

Wu, F. (2002), "New partners or old brothers? GONGOs in transitional environmental advocacy in China", *China Environment Series*, 5, 45–58

Wynne, B. (1992), "Uncertainty and environmental learning: Reconceiving science and policy in the preventative paradigm", *Global Environmental Change*, 2(2), 111–127

Wynne, B. (1996), "May the sheep safely graze? A reflexive view of the expert-lay knowledge divide". In S. Lash, B. Szerszynski, and B. Wynne (Eds.), *Risk, Environment & Modernity: Towards a New Ecology*, Cambridge, U.K.: Polity, pp. 44–83

Xie, L. (2007), *Environmental Activism in Urban China: The Role of Personal Networks*, Wageningen: Wageningen University (Ph.D. dissertation)

Xie, L., and A. P. J. Mol (2006), "The role of *guanxi* in the emerging environmental movement in China". In A. M. McCright and T. N. Clark et al. (Eds.), *Community and Ecology*, Amsterdam: Elsevier/JAI, pp. 269–292

Yang, G. (2005), "Environmental NGOs and institutional dynamics in China", *China Quarterly*, 181, 47–66

Yang, G. (2007), "How do Chinese civic associations respond to the Internet: Findings from a survey", *China Quarterly*, 189, 122–143

Yearley, S. (1991), *The Green Case: A Sociology of Environmental Issues, Arguments and Politics*, London: Harper Collins

Yearley, S., S. Cinderby, J. Forrester, P. Bailey and P. Rosen (2003), "Participatory modelling and the local governance of the politics of UK air-pollution: A three-city case study", *Environmental Values*, 12, 247–262

Yelvington, S. (1999), http://G/cd/keynote/yelving.htm (accessed October 2004)

Yong, J. S. L. (2003), *E-Government in Asia: Enabling Public Service Innovation in the 21st Century*, Singapore: Times Editions

York, R., and E. A. Rosa (2003), "Key challenges to ecological modernization theory", *Organization and Environment*, 16(3), 273–287

Young, I. M. (1990), "The ideal of community and the politics of difference", in L. Nicholson (ed.) *Feminism/Postmodernism* New York: Routledge, pp. 300–323

Young, O. (1994), *International Governance: Protecting the Environment in a Stateless Society*, Ithaca, N.Y.: Cornell University Press

Young, S. (1995), "Participation – out with the old, in with the new?" *Town and Country Planning*, April, 110–112

Zald, M. N., and J. D. McCarthy (1979), *The Dynamics of Social Movements: Resource Mobilization, Social Control and Tactics*, Cambridge, U.K.: Winthrop

Zavestoski, S. and S. W. Shulman (2002), "The Internet and Environmental Decision Making: An Introduction", *Organization & Environment*, 15, 3, pp. 323–327

Zook, M. A. (2001), "Old hierarchies or new networks of centrality? The global geography of the Internet content market", *American Behavioral Scientist*, 44(10), 1679–1696

Index